C000193826

BRUNEL *in* SOUTH WALES

BRUNEL
in
SOUTH WALES

VOLUME I

IN TREVITHICK'S TRACKS

STEPHEN K. JONES

Stephen K Jones

TEMPUS

In memory of my father and uncles:
Kedwyn Hathaway Jones,
Lawrence Douglas Jones and
Melvyn James Jones.

Cover illustration:
Goitre Coed Viaduct, 1841, by Penry Williams.
(Courtesy of the Elton Collection of the Ironbridge Gorge Museums)

Taff Vale Railway uniform button and the £2 Trevithick coin issued in 2004. (SKJ photograph)

First published 2005
Reprinted 2006

Tempus Publishing Ltd
The Mill, Brimscombe Port
Stroud, Gloucestershire GL5 2QG
www.tempus-publishing.com

© Stephen K. Jones, 2005

The right of Stephen K. Jones to be identified as the Author of this work has been asserted by him in accordance with the Copyrights, Designs and Patents Act 1988.

All rights reserved. No part of this book may be reprinted or reproduced or utilised in any form or by any electronic, mechanical or other means, now known or hereafter invented, including photocopying and recording, or in any information storage or retrieval system, without the permission in writing from the Publishers.

British Library Cataloguing in Publication Data.
A catalogue record for this book is available from the British Library.

ISBN 0 7524 3236 2

Typesetting and origination by Tempus Publishing.
Printed and bound in Great Britain.

Contents

In Trevithick's Tracks

From the world's first railway locomotive to the locomotive railway in Wales – an introduction to Brunel in South Wales.

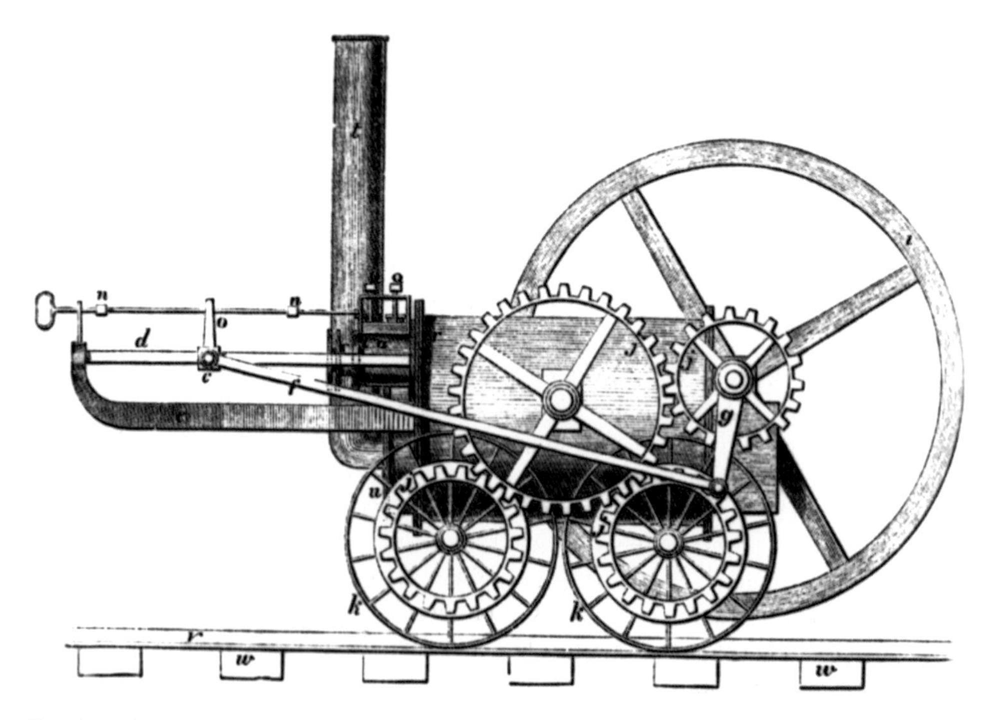

Penydarren locomotive

Series Foreword

Of all the great British engineers of the last two centuries, the best known and most popular for the present-day public is I.K. Brunel, who was born in 1806 and died in 1859 at the age of fifty-three. His popularity is partly due to the success of a brilliant but informal public relations exercise whereby Brunel managed to get himself photographed, in the infancy of photographic portraiture, standing in a confident posture in front of the giant chains used to restrain the launch of the PSS *Great Eastern*. The photograph by Robert Howlett has become an iconic image, frequently reproduced to represent Victorian enterprise and engineering prowess, so that Brunel's face has become familiar to a wide audience.

But it is not simply a matter of appearances: Brunel did not merely look like a dominant engineer. His career, although cut short by his premature death, was creative and spectacular. He masterminded, in the Great Western Railway, the establishment of one of the main national railway systems, which with its smooth gradients and distinctive broad-gauge track, set new standards of high-speed passenger transport and excellence of service. Then he went on to build three massively innovative steamships, designed explicitly to promote long-distance oceanic passenger traffic, by taking huge steps in enlarging the size of ocean-going vessels and introducing new techniques of construction and propulsion. In the course of these projects he experimented with iron structures to produce some outstanding bridge designs and undertook many lesser commissions to build docks and waterworks, to improve drainage and design hospitals. He was, from any point of view, an extraordinarily imaginative and successful engineer.

It was originally intended that the GWR would link Bristol with London, but under Brunel's inspiration it was rapidly extended through subsidiary lines to Exeter and on into Cornwall, and branches went off to the north, to Oxford and Worcester and beyond, and to the south, to Weymouth and elsewhere. It was not long before Brunel began to explore the possibility of extending the system into South Wales. The iron and coal traffic through Cardiff and Newport was an attractive incentive for business investors, and beyond these there was the prospect of opening up a high-speed rail link to Milford Haven and on to Ireland, providing an alternative to the main route to Dublin through North Wales and Holyhead. Brunel had already made the acquaintance of the ironmasters of Dowlais and Merthyr through his commission to build the

Taff Vale Railway, and he immediately established relations of mutual respect and collaboration with them because his ships and railways created a tremendous demand for their products. The snag, so far as the GWR was concerned, was that the Severn Estuary imposed a substantial barrier to communication, so that any operation in South Wales would be in danger of becoming a limb detached from the main body of the network. Brunel knew that the estuary would soon be either bridged or tunnelled, and both connections were achieved within a few decades of his death, but in his lifetime it was necessary either to take a ferry, or travel the long way round via Gloucester.

The link between Bristol and Gloucester was made almost immediately but was then complicated by competition with the Midland Railway, although the GWR retained access to Gloucester from Swindon and Stroud, and so felt able to promote the Gloucester–Forest of Dean line. Connection with the South Wales Railway, which was also promoted in the 1840s, had to await the completion of Brunel's novel bridge at Chepstow. Although nominally independent of the GWR, these enterprises were supported by resources from the parent company and, more to the point, they used the same engineer. From Brunel's point of view they were all part of the network for which he was responsible and he devoted considerable attention to the South Wales extension of this system. Not only did the Wye crossing at Chepstow allow him to develop his distinctive ideas of wrought-iron trusses as a means of establishing wide spans, but elsewhere on the line through to Pembrokeshire he found opportunities to devise ingenious structures in wood and masonry, as well as iron. Even on the Taff Vale Railway, which was destined to be a standard-gauge line and did not link directly with the broad-gauge main line, Brunel created the innovative viaduct at Goitre Coed.

Brunel thus made his mark on the landscape of South Wales, and many signs of his masterful presence survive and are accessible to the careful observer. It is strange, therefore, that so little attention has been given to this aspect of the engineer's career. Brunel studies have tended to be centred on Bristol, rightly seeing this city as the key to so many features of his work, even though he remained throughout his life a stubbornly metropolitan man and never lived in Bristol. His home was above his office in Duke Street, Westminster, and when he did get round to investing in a country estate, it was not in the Bristol region but at Watcombe, near Torquay. With this London–Bristol–West Country axis in mind, it has been easy to minimise the importance of Brunel's work in South Wales. But at last it is possible to welcome a study that comprehensively makes good this omission.

It would be hard to imagine anybody better qualified to write on Brunel in South Wales than Stephen K. Jones. He is, in the first place, a native Welshman with a keen sense of the role of Wales in modern industrial history. He has also been a long-term admirer of I.K. Brunel, and has devoted an enormous amount of time and energy to pursuing his interest in the engineer. I met Steve first of all as a member of the Brunel Society, which is now sadly defunct, but which did sterling work in arousing interest in the life of Brunel during the 1960s and 1970s. We collaborated fruitfully on two studies of Brunel themes, the Balmoral Bridge and the Crystal Palace Water Towers, both of which have been published in *Industrial Archaeology Review*. I have come to have a hearty respect for his assiduity as a researcher and for his skill in assembling the results of his enquiries into an informative text. His research has always been extremely thorough and he has never failed to pursue any opportunity to discover information on the subject in hand.

This present study is the result of many years of application to the hard work of documentary research, of taking photographs to record standing structures and other features, and of shaping it up into lectures and articles. I have heard Steve giving admirable illustrated lectures on Brunel in South Wales, and I am delighted to be able to welcome this complete version of his study. The first volume brilliantly sets the scene for Brunel's activity in South Wales, focussing on his design and construction of the Taff Vale Railway. Subsequent volumes deal with the main South Wales Railway, other broad-gauge lines and the maritime connections with South Wales. I commend the book to all railway historians and Brunel fans. With the 200th anniversary of Brunel's birth coming up in 2006, it is most appropriate that this book should appear now, making such an important and authoritative contribution to the celebration of that event.

R. Angus Buchanan
University of Bath
2005

Looking forward; Lady Cynthia Gladwyn, the great-granddaughter of Isambard Kingdom Brunel. She is seen here at Sydney Gardens, Bath, following the unveiling of a Brunel Society plaque commemorating the opening of the Bristol to Bath section of the GWR on 31 August 1977. (SKJ photograph)

Foreword To Vol.1 In Trevithick's Tracks

The industrial history of South Wales may appear, from a superficial standpoint, to be a relatively simple one. Iron ore was converted into iron, coal was mined, and the two products were transported to the coast for onward shipment elsewhere for consumption or further processing. In reality, this history is a much more complex one in that the main phases of iron and coal production did not coincide, while the change to steel production brought about its own geographical changes.

The mechanisms of transport involved, initially, rough tracks, followed by canals and their associated tramroads. In turn, the latter metamorphosed into some of the earliest railways, which were almost always in a state of vehement competition with each other. This transition in South Wales also witnessed the world's first operational railway locomotive. There were also the individual ironworks, all in competition with each other, but showing a degree of collaboration when workers' demands were to be resisted. Superimposed on all these elements were the ironmasters themselves with their individual and, usually, sharply defined personalities. The coal masters formed an equally idiosyncratic group but came rather later.

The resulting complex evolution did not, of course, take place in isolation from the rest of Britain. Most of the ironmasters had their origins outside Wales, as did many of the consulting and contracting engineers. This overall picture of products, places of production and extraction, owners and engineers, and mode of transport, is thus an extraordinarily involved one. It may be argued that, at the centre of this web, lay the Penydarren Tramroad, together with the outstanding personalities of Richard Trevithick and I.K. Brunel.

It is important to understand the interrelationships in the evolution of the complex structure of this industrial region. Such an explicit presentation has been provided by the author of this book. The overall picture, which he unfolds logically and clearly, is based upon many years of study and the accumulation of much detailed information.

This evolutionary presentation also provides the reader with the background to, and the operational construction of, the Taff Vale Railway, and also to the circumstances leading up to Brunel's role in the realisation of that project. This railway was central to the development of the iron and coal industries of the Taff Valley, and was the precursor of what became a maze of railways in South Wales.

While there have been various publications over the years dealing with some of the individual aspects referred to above, this book provides an integrated presentation, setting out the circumstances and implementation of one of the greatest industrial stories of the world to date.

Stuart Owen-Jones,
2005

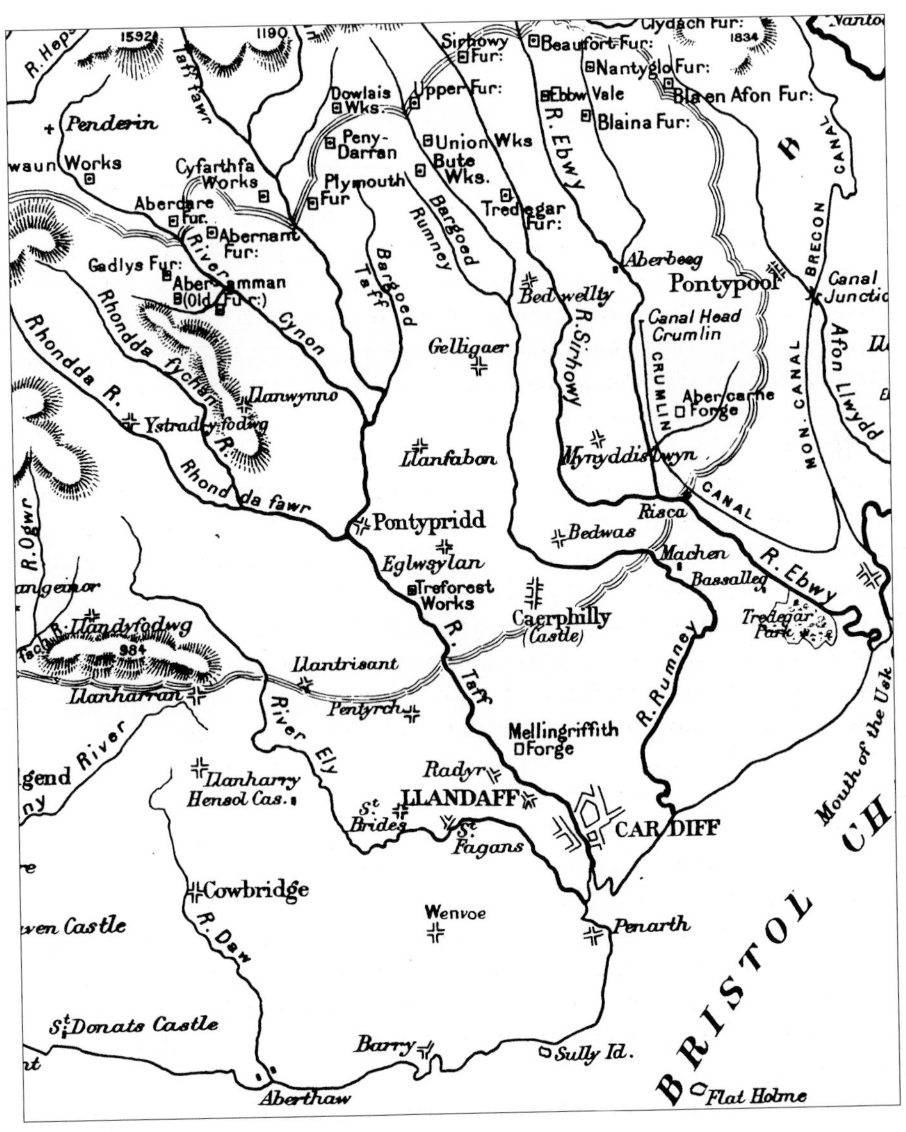

Adapted from John Lloyd's map illustrating 'The Early History of the Old South Wales Iron Works, 1760 to 1860'.

Preface And Acknowledgements

Brunel in South Wales first took shape as a slide lecture to the Brunel Society some thirty years ago and has been a work-in-progress ever since. What became obvious from my initial research was that South Wales had been largely overlooked as an area of interest relating to the works and activities of Isambard Kingdom Brunel (1806-59), yet it was important in terms of engineering landmarks at virtually every stage of Brunel's career. Writing this book is an attempt to redress this and illustrate the achievements and the legacy of the engineer, covering his works against a background of social and industrial history.

In Trevithick's Tracks provides an introduction to *Brunel in South Wales*, from the world's first railway locomotive to the first locomotive railway in Wales, the latter being Brunel's first commission in South Wales, the Taff Vale Railway. It first received his attention in 1834 and in many ways represents a transitional stage for Brunel the railway engineer, and was to be his only narrow- or standard-gauge (4ft 8½in) line in this country. It was promoted as a railway from Merthyr Tydfil to Cardiff, and as such can be seen as the final stage in transport improvements brought about by the iron industry of Merthyr Tydfil. One such improvement was the Merthyr Tramroad of 1802, on which Richard Trevithick was to make his historic journey in 1804, the first steam locomotive to run on rails in the world. It was a revolutionary innovation that was to be, sadly, developed elsewhere.

Thirty years later, Isambard Kingdom Brunel was approached to engineer a locomotive railway, bringing the development of Trevithick's ideas back to its birthplace. The Taff Vale Railway represents the logical evolution of improvements in transport and was the last stage as far as the South Wales iron industry was concerned in serving their needs and greatly accelerating the growth of another industry – coal.

Isambard Kingdom Brunel had asked his friend and professional rival, Robert Stephenson, if he could recommend any contractors capable to tender for work on what was to be his first railway in Wales. 'A crooked line down in Wales' was how Stephenson was to describe Brunel's Taff Vale Railway to a potential contractor! It is ironic that although being described (albeit in jest) as a crooked line, the Taff Vale Railway was one of the first for which a mathematical formula for calculating the curvature of the line was developed. The calculations were made by an assistant engineer but were to catch Brunel out, in terms of his decision for the gauge of the line and the later ramifications of the Gauge Commission.

The Taff Vale Railway was also one of Brunel's first railway commissions, coming soon after the GWR. With such demands on his time it might be assumed that he would employ existing standards of engineering design and construction on the TVR, particularly in view of the 'Stephenson' features, including the gauge, on the finished line. But even here there were major engineering works that showed his innovative approach, such as the crossing of the river Taff, which had to be made on the skew. Goitre Coed Viaduct was, however, no conventional skew arch but a totally Brunelian solution of a viaduct, with six 50ft spans springing from octagon piers, the side of the octagon being parallel to the axis of the river. This is just one example of the unique and innovative engineering solutions that make up Brunel's legacy in South Wales.

Communications and Coal will be the basis for the next volume covering mail routes, main lines and mineral railways, showing that Brunel's engineering philosophy was now clearly defined with his return as the engineer of the South Wales Railway. On this broad-gauge (7ft ¼in) line, several novel engineering ideas were worked out, such as bowstring girders and tubular suspension spans. The latter were intended for bridging the river Severn, a daring design that did not materialise, but one which would later take shape at Chepstow over the Wye and Saltash across the Tamar.

Extensive use was also made of timber viaducts on the South Wales Railway, including his largest timber work and the only surviving example of a bridge, supported on the original timber piles and piers, to carry main line traffic to this day. Broad-gauge mineral lines followed, such as the Vale of Neath Railway, which the Stephensons said could not be built because of the gradients involved. On the Vale of Neath Railway, Brunel was to introduce a container system to minimise coal breakage from the coal mine to ship.

Links with Leviathans will cover the maritime dimension, of ships and shipping facilities, to Brunel's work in South Wales. The final volume will cover these associations and the links with leviathans of the sea, such as his great iron steamship, the PSS *Great Eastern*. Brunel selected Milford Haven as a home port and another South Wales port saw the last commercial voyage of his SS *Great Britain* steamship. Loaded with Welsh coal, the SS *Great Britain* sailed from Penarth, a dock associated with the later work of Brunel's son, Henry Marc (who was later responsible for the largest Welsh dock at Barry). Other docks include Briton Ferry which he designed to handle the output of the Vale of Neath Railway, and another broad-gauge line, the South Wales Mineral Railway.

Brunel made extensive use of local industry. The great Dowlais ironworks at Merthyr supplied rails for several of his lines and ironwork for the Hungerford and Clifton Suspension Bridges. The PSS *Great Eastern* was equipped with chain cable made by Brown Lenox of Pontypridd (used as a backdrop to the most famous photograph of the engineer). One of his last projects was a steam railway ferry across the Severn, another unique feature of *Brunel in South Wales*.

In terms of place-name spellindgs and so on, I have attempted to keep to the original spellings as used at the time; however, in the case of Goitre Coed Viaduct, where Brunel refers to it as 'Godre y Coed', TVR contracts as 'Goche-y-coed' and Samuel Downing as 'Goed-re-Coed', I have used the Ordnance Survey spelling. The nearby tunnel was also known as 'Penlock's' tunnel, and by Lady Charlotte Guest as 'Cefn Glas'.

How *Brunel in South Wales* first took shape has already been mentioned, and I wish to record my appreciation to Mike Williams and Keith Hickman of the Brunel Society and to Philip Riden of Cardiff University for encouraging me to develop the material into an university extra-mural course. Researching the subject over the years has meant that there is a debt of gratitude owed to a number of individuals that can no longer be thanked personally; Eric R. Mountford; John G. James; Ray Bowen; John A. Owen; Owen Gibbs; Ken Holloway and Bill Hamlin being foremost amongst these. In completing this work I am indebted to Angus Buchanan; Colin Chapman; Stuart Owen-Jones; Paul Reynolds and Stephen Rowson for their constructive suggestions and support. Particular thanks for help with information and photographs go to Tom Morgan; John Minnis; David Weston and the Hon. Sir William McAlpine; Brian L. James; Alan George; Geoff Matsell; Derek Portman; Anne Gardner; Nigel Billingham; Bill Daniels; Gary Greenslade; W.D. (Bill) Edwards and the Dr G. Sauer Co. Ltd.

For access to archives and illustrations I would like to acknowledge the assistance given by the Central Reference Library, Cardiff; the Public Record Office, Kew; the National Museum and Galleries of Wales (Welsh Industrial & Maritime Museum collection); the National Railway Museum, York; the University of Cambridge Air Photograph Library; Susan Edwards of the Glamorgan Record Office; Michael Richardson and Hannah Lowery of the University of Bristol (University Library Brunel Collection); Carol Morgan and Mike Chrimes of the Institution of Civil Engineers (library and archives); Ron Cox of the University of Dublin; Michael Vanns and John Powell of the Ironbridge Gorge Museums; Angela Broome of the Royal Institution of Cornwall; William Troughton of the National Library of Wales; Keith Gilroy of the Northumberland Record Office; Rhys Griffith of the Herefordshire Record Office; Tony Jukes and Robin Williams of Oxford House Industrial Archaeology Socity; Lawrence Smith-Higgins of the Patent Office, Newport; Alastair Warrington and Alan Evans of Networkrail Great Western; Meurig Morgan of Cyfarthfa Castle Museum, Merthyr Tydfil and Caroline Jacob of the Merthyr Tydfil Reference Library. Special thanks go to Owen Eardley for his help in all matters visual. My sincere apologies go to any whom I have inadvertently omitted. At Tempus Publishing I am grateful to Wendy Tse, my first contact in connection with the project, and to Amy Rigg, Nicola Sweet and Campbell McCutcheon for their patience and help in completing it. Above all, my greatest debt is to those who have put up with the subject on almost a daily basis, namely my wife, Siân, without whose support there would be no book, and our daughters, Kathryn and Bethan.

All errors are regretted and entirely my own.

Stephen K. Jones,
2005

1
An Engineering Dynasty
'Quand je serai grand, j'irai voir ce pays-là'[12]

The Brunel engineering dynasty started in France with the birth of Marc Isambard Brunel, who was born on the 25 April 1769 at Hacqueville, near Gisors in Normandy. His background, in terms of family tradition, was firmly entrenched in the Normandy line of yeomanry farming, and Marc, as the youngest son, was destined for the church. In training for this role he was sent to the College of Gisors at the age of eight, where he was taught Latin and Greek, subjects for which he had no inclination, instead preferring drawing and mathematics.

At the age of eleven, Marc was sent to the Seminary of St. Nicaise at Rouen for further ecclesiastical training but, by this time, he was determined to follow his own career path and qualify for the navy. The Principal of St Nicaise realised that Marc was not committed to the calling and persuaded Marc's father, Jean Charles, not to force the boy into the traditional vocation. Jean Charles eventually relented and allowed him to follow his heart's desire. Marc was sent to study hydrography at Rouen under Monsieur Dulague, and stayed with a family friend. This was Francois Carpentier, a retired sea captain who also served as the American consul at Rouen. Dulague was amazed at the speed at which Marc was able to absorb the propositions of Euclid and, after only his third lesson in trigonometry, the young student was making an instrument with which to measure the height of Rouen Cathedral.

It was during this period that the young Brunel saw two cast-iron cylinders being unloaded on the quay at Rouen. His inquisitiveness about these strange devices led to the answer that they formed part of a new 'fire' engine for pumping water, and had been made in England. 'Ah!' exclaimed Marc; 'quand je serai grand, j'irai voir ce pays-là.'[13] This experience was to stay with him and make him determined to one day visit the country from which such machinery could originate.

With the help of Dulague, Marc obtained a commission from the Minister of Marine as a 'Volontaire d'honneur' on the corvette Le Maréchal. He made his first voyage in 1786, to the West Indies, and continued to serve for a further six years, gaining the nickname 'Marquis' (Marc I) from his fellow sailors.[14] He made his own quadrants for use at sea, using his second improved model for most of his voyages.

In 1792 his ship was paid off and Marc went to Paris, but his openly expressed loyalist opinions soon led to his departure from the city.

Returning to Rouen, he took refuge with his old patron Capt. Carpentier, who offered to obtain a passport for him to leave France. Staying with the family at this time was a young English woman called Sophia Kingdom, who had been sent to learn French. She was the orphaned daughter of William Kingdom, a Plymouth naval contractor. Against the revolutionary background Marc and Sophia formed an attachment that was to become a lasting relationship. There was little time to spend together as Marc had to leave France in a hurry on 7 July 1793, as he had found a berth on the American vessel *Liberty*. Soon after the start of the voyage, the ship was stopped by a French frigate with orders to arrest royalist sympathisers. At this moment, Marc realised that the passport Capt. Carpentier had secured for him was missing. In the short time available, and with only the resources he could find, Marc was able to forge another passport which passed scrutiny by the boarding party.

On 6 September 1793, Marc arrived at his destination:'This day in 1793 I landed at New York as an exile from my country. Luckily with ample means for a bachelor to wait for some opportunity of doing something.'[15]

The opportunity presented was in the form of surveying and civil engineering. His first engagement was on the survey of a large tract of land near Lake Ontario, followed by another survey for a proposed canal from the Hudson River to Lake Champlain. This project had been originally in the hands of another French emigré, but the difficulties encountered led to the commission being offered to Marc. Having completed these commissions, his fame spread and he undertook architectural work, competing in the competition held for the design of a new House of Assembly in Washington. His design was well received but ruled out on the grounds of cost. The design was modified by Marc for the Palace Theatre in the Bowery district of New York, which was built, but burnt down in 1821.[16]

Relief bust of Marc Isambard Brunel on the Hacqueville Brunel monument, Normandy, France. (SKJ photograph)

Brunel farmhouse in the centre of Hacqueville, Normandy. (SKJ photograph)

Now an American citizen, Marc secured the post of Chief Engineer of New York, work which included designing a cannon foundry and arsenal as well as improving the defences of Long Island and Staten Island. The opportunities presented by the New World should have kept an ambitious engineer busy, but two thoughts kept calling his attention to Europe. The first was his intention to visit England that had been formed on the quayside at Rouen, and the second, his desire to renew his friendship with Sophia Kingdom.

An opportunity presented itself in 1798 when Marc dined with the prominent American statesman Alexander Hamilton (1757-1804) and met a fellow exile, M. Delabigarre. During the dinner, conversation turned to Britain's victories at sea and how the expansion of the Royal Navy was threatened. This was caused by the slow production of ships' blocks, an area that Delabigarre appeared to have expert knowledge of, and which was causing delays in rigging and commissioning ships. From this conversation, Marc evolved the idea of mass producing ships' blocks by the use of specially designed machinery. He would later say that the first idea of the block machinery came to him when dining with Hamilton. The thoughts were further developed; he later found himself cutting certain initials into the bark of a tree, when it struck him that the curve of one of the letters could lend itself to the process: '… this is it' he cried, 'my pulley shall have this curve'.[17] Resolved to submit his ideas to the British Government, Marc obtained an introduction from Alexander Hamilton to see the First Lord of the Admiralty, Earl Spencer of Althorp. As First Lord of the Admiralty under Pitt, Earl Spencer greatly improved the administration of the navy and was responsible for putting down the mutinies that had broken out

at the Nore and Spithead. It was a good introduction for Marc and served beyond the work in hand as he was to remain on friendly terms with the Spencers, and make several visits to Althorp and its magnificent library.

In March 1799 Marc arrived at Falmouth and from there he went straight to see his Sophia. They were married within the year. In order to promote his ideas on blockmaking he decided to have working models made, and approached Henry Maudslay who was to expand his business as Maudslay, Sons & Field, which would become one of the most famous engineering firms of the nineteenth century. Maudslay's contribution was significant and led to the construction of iron machinery that not only heralded a new age of mass production, but stood out as exemplars of precision and reliability for the machine tools of the future. In her biography, Celia Brunel Noble recalls that her great-grandfather had brought with him: '… some small means and many great ideas.'[18] But there were also many obstacles, one being that Marc was French and therefore regarded with great suspicion in the aftermath of the French Revolution and Napoleon's rise to power. Conservatism, in terms of the existing methods employed and the entrenched views of manufacturers, was another obstacle. Marc approached the leading supplier of ships' blocks to the navy, Taylors of Southampton, through his future brother-in-law. Taylors flatly rejected his offer to supply new machinery. The Spencer connection was then used to bring his ideas to the attention of Brigadier General, Sir Samuel Bentham, the Inspector-General of Naval works. Bentham had carried out many improvements in the naval dockyards and had been planning to introduce machinery to speed up block-making when Taylor's contract expired. Greatly impressed by Marc's scheme, which was one

Opposite page: *The Haqueville Brunel monument, Normandy. (SKJ photograph)*

Right: *The inscription on the Haqueville Brunel monument, Normandy. (SKJ photograph)*

of the first examples of mass production, Bentham recommended its adoption by the Admiralty.

The proposal was accepted, with the installation of block-making machinery at Portsmouth, and the contract for machinery going to Maudslays. Work began in May 1803 and was completed three years later, with Marc directing the erection of machinery at the dockyard. In order to be near his work, Marc moved to Portsmouth, and he and Sophia settled into a small house at Portsea. Here, two daughters and a son were born to the Brunels. In 1806 Marc made the following entry in his journal: 'On 9 April, and at five minutes before one o'clock in the morning, my dear Sophia was brought to bed of a boy.'[19] The boy was christened Isambard after his father, and Kingdom after his mother.

The start of the year 1806 was marked by Nelson's funeral and, a few days later, by the death of the Prime Minister; William Pitt, who was succeeded by Lord Grenville. Napoleon was to launch his Continental System at Berlin in 1806, with the intention of sealing off Britain from Europe and bringing her to her knees. Britain's sea lanes, however, remained open but cross-channel trade was severely affected. The decline affected merchant trade on these routes with the knock-on affect of the closure of several private shipyards. However, in order to keep Britain's sea lanes open, particularly to new markets or the reopening of old markets in the Baltic, America, Portugal and the near East, the Royal Navy had to be kept up strength and be fully equipped.

Marc continued to develop his ideas for mechanisation, with several machines being designed and patented for sawing and handling timber. Further orders were placed by the Admiralty to equip a sawmill at Chatham Dockyard. At Battersea, Marc set up a sawmill of his own, alongside which he was later to build another factory, this time for mass producing military boots, the impetus for which had come from Marc seeing the wretched state of troops returning from Corunna. There was much jubilation in 1814, following the fall of Paris and Napoleon's abdication and exile to Elba. The visit of the allied sovereigns to Britain was a major event that was covered by newspapers such as *The Times* in their 11 June edition. The dignitaries found time away from the celebrations and negotiations to see Portsmouth's naval works and: '… that unequalled system of machinery for making ships' blocks, which they thought was of itself worth coming to Portsmouth to see.'[20]

The Government agreed to buy all the boots manufactured in Marc's factory, but disaster struck on 30 August 1814 when fire broke out in the sawmill. Within a few hours the sawmills were completely destroyed and the blaze had spread to the boot factory. Although the sawmill was eventually rebuilt, the fire was a considerable setback and, with the peace that followed victory at Waterloo, Marc was left with a large stock of boots that the army refused to buy. These circumstances were to lead ultimately to Marc's imprisonment for debt, and he spent two months in the King's Bench prison. Eventually, the Government, prompted by the Duke of Wellington, paid his debts on the condition that he would remain in England.

Following his release, Marc's engineering activities ranged over a wide field. He was involved in steam navigation, invented a knitting machine, improved printing-plates and designed several bridges. Two of these were suspension bridges, designed for the French Government, and erected on the Ile de Bourbon. There were more sawmill commissions, for Trinidad and Berbice, and swing bridge and landing stage designs for Liverpool docks. Other projects included experiments with 'gaz' engines and his last and most ambitious work, the Thames Tunnel.

A sub-aqueous tunnel under the Thames was not an original idea in itself. In 1798 Ralph Dodd (*c*.1780-1837) put forward a proposal for a tunnel, following his earlier paper proposing a tunnel under the river Tyne. An Act was obtained and work began in 1799 on the tunnel which was to run from Gravesend to Tilbury, but the destruction of the pumping engine by fire at the end of 1802 brought all activity to a standstill.[21] With funds exhausted, the company was wound up, but it did not stop another tunnel project being put forward by the Cornish engineer, Robert Vazie, who proposed a tunnel from Rotherhithe to Limehouse. This led to the formation of the Thames Archway Co., which was incorporated by an Act of Parliament in 1805. Construction of this tunnel was no easier, and early work was fraught with difficulties and floodings. Eventually, another Cornish engineer, Richard Trevithick (1771-1833), was called in by Vazie on the advice of Davies Giddy (1767-1839). He was later to be known as Gilbert, a name he assumed in 1814 for family reasons, and for consistency will be referred to from hereon by this name. Gilbert was a fellow Cornishman and mentor to Trevithick, and recommended him when the engineers, John Rennie and William Chapman, could not agree on the way forward. Gilbert had sat on the parliamentary committee to consider the Thames Archway Co.'s Bill when it went through Parliament, and Trevithick greatly appreciated his recommendation.[22]

What is of interest about the personalities involved here is that Trevithick had come to the tunnel just a few years after his pioneering work in South Wales (see chapter three), at the recommendation of Gilbert, who was also to play a part in the procedure by which Isambard Kingdom Brunel was to secure his first independent commission, while Dodd was to be later involved with George Stephenson's first locomotive engine at Killingworth. One of Trevithick's first actions was to install one of his high pressure, or 'strong steam', engines to drain the tunnel works. His plan was to dig a driftway or pilot heading first, which, on reaching the opposite side of the river, would be opened out to the full tunnel size. Shifting sands and high tides caused an inundation into the driftway on 26 January 1808. Trevithick arranged clay bags to be deposited from boats to plug the gap in the river bed, and work resumed. However, a further inundation just over a week later was to mark the end of Trevithick's tunnelling endeavours, despite further deposits of clay bags. Trevithick even came up with a novel idea for completing the tunnel using cast-iron tunnel sections laid in a trench in the river bed which would be excavated using a series of coffer dams. Thames Archway Co. would have none of it and invited alternative ideas, to be judged in a competition by the eminent mathematician, Dr Charles Hutton, and the well-known engineer, William Jessop.[23] Suggestions poured in from all sectors of the public and were carefully scrutinised, but the judges concluded '… under the circumstances which have been so clearly presented to us, we consider that an underground tunnel, which would be useful to the public and beneficial to the adventurers, is impracticable.'[24] This statement was to be eventually disproved by Marc although his tunnel would also attract comments and suggestions when it ran into difficulties almost twenty years later. Looking back on this work, Marc was to relate to Richard Trevithick's biographer (his son Francis) that, '… the plans and papers on the driftway left by Trevithick had given him the highest opinion of his character and genius.'[25]

In his tunnel endeavour, Marc was to follow other engineers' attempts but he set out to accomplish this task through an innovative approach: the use of a tunnelling shield. This idea, which would protect the miners working at the face, had been conceived by Marc while watching the destructive work of the shipworm (*Teredo navalis*) in a piece of ship's timber. Marc was approaching the height of his engineering career when he embarked on the project that was to dominate the rest of his working life, a period which marked the start of his son's active involvement in engineering. The Thames Tunnel overlapped both careers. For Marc it was to be his last major work, while for Isambard it was the beginning.

Marc had taught young Isambard the techniques of mechanical drawing he had learnt, such as the representation of three-dimensional objects in a two-dimensional plane. Such techniques were relatively new to British engineers, being French military secrets for some thirty years after having been evolved by Gaspard Monge of Mezieres in 1765.[26] Isambard inherited his father's flair for mechanics and quickly assimilated knowledge. His skill at drawing was displayed from the age of four and, by the age of six, he had mastered Euclid. Early formal education was at Dr Morrell's boarding school at Hove and even there he showed his growing perception of engineering matters by correctly predicting the collapse of a new building opposite the

Lithograph of the Thames Tunnel, showing the staircase access from the street. (Courtesy of Dr G. Sauer Co. Ltd)

school. Next, he was sent to France for a period of two years where he was taught at the College of Charlemagne, better known as the Lycée Henri-Quatre, and the school of Monsieur Massin. Isambard's school report from the Institution de M. Massin lists prizes in mathematics, French language and drawing, and gives short summaries of his performance in nine out of thirteen subjects.[27]

During this period, Isambard also worked under Abraham-Louis Breguet (1747-1823) in Paris. This was seen as an opportunity for practical training with the man regarded as the supreme craftsman in the field of chronometers, watches and scientific instruments. Indeed, Breguet was developing his 'chronometre a doubles secondes' (observation chronometer) which anticipated the modern chronograph, and his 'inking chronograph' at around the same time.[28] Marc had chosen Breguet because he felt that his son's view of engineering should not be confined to a narrow base. The necessary exacting standards of the expert instrument maker were to be adopted by Brunel in his engineering career.

With his French education and training, Isambard returned to London in 1822 and, at the age of sixteen, went to work with his father, whom he assisted in his small office at No.29 Poultry. As part of this stage of his apprenticeship, he spent a considerable amount of time and gained much useful experience in engineering workshops such as the Lambeth works of Maudslay, Sons & Field. Henry Maudslay had become a close friend of his father and his workshop presented an opportunity to witness the great strides Britain was making in industrial development. Marc was to express his confidence in Isambard's mathematical ability on many occasions, and even in a professional capacity. On one occasion in 1825, when the nineteen-year-old Isambard was representing his interests, Marc stated in a letter that he was not afraid of him making: '... any calculations that might be necessary to prove the practicability of my plan ...'[29]

It was to come as no surprise then that in 1825 Isambard would take an active part in his father's most ambitious project, the Thames Tunnel. Appointed by his father to assist the resident engineer George Armstrong, his role was to increase in importance when he was made temporary resident engineer during George Armstrong's illness and convalescence. Isambard urged the construction on and the heading extended, on average, 8ft a week in his first two months. Early in August 1826, George Armstrong resigned and Isambard was promoted to resident engineer, but it appears that Marc did not consider any other candidates for the post.

Work progressed despite occasional flooding and Isambard spent long periods of time underground. On one occasion in September 1826, he worked for 96 consecutive hours with only a few snatches of sleep. During the first major flooding of the tunnel on 18 May 1827, Isambard was to save a workman from drowning by lowering himself into the flooded tunnel shaft and bringing the man to the surface with the help of his assistant, William Gravatt (1806-66). This selfless act of bravery

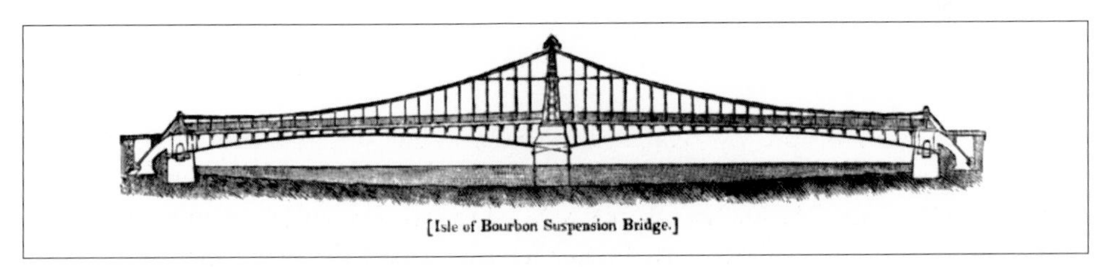

Marc Brunel was commissioned by the French Government to design two suspension bridges for Ile de Bourbon in the Indian Ocean (each of 132ft span). (SKJ collection)

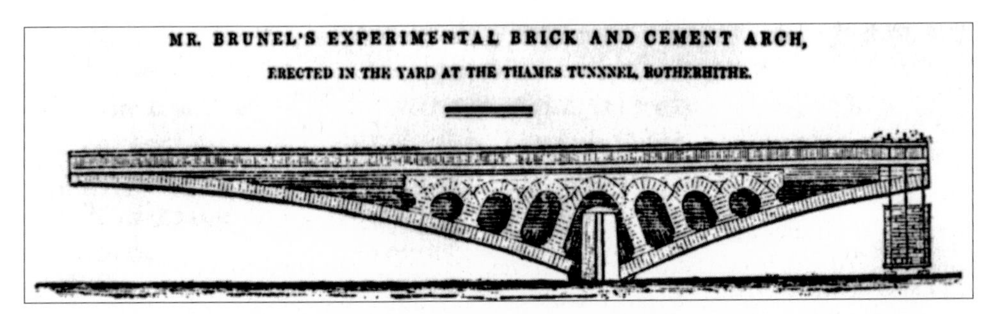

Experiments to test cement bonds for brick arches. Isambard was to be actively involved in these experiments and the practical experience and knowledge gained from both types of bridge construction was to be of great benefit to him at the start of his career. (SKJ collection)

resulted in them both being awarded a Royal Humane Society lifesaving medal on 5 March 1828.[30] In the meantime, the river broke in again on 12 January 1828, and Isambard narrowly avoided drowning. Although seriously injured, he refused to leave the works until he had directed the dumping of clay over the path of the tunnel. Further dumping filled the cavity in the river bed above the workings but it was not until the beginning of April 1828 that the tunnel could be pumped dry. The cost of making good this, the second major flooding, exhausted the funds of the Thames Tunnel Co. and, in August of the same year, all work ceased and the tunnel shield was bricked in.

Whenever there was a serious inundation in the tunnel, the press and, in particular, engineering journals such as the *Mechanics Magazine*, would carry numerous letters and editorial pieces offering advice on how disaster could have been avoided and how work on the Thames Tunnel could be resumed. Several decades later Isambard Kingdom Brunel was to receive similar treatment over the launch of the *Great Eastern*. In 1827, Thomas Deakin of the Blaenavon ironworks commented in a letter that the tunnel had not been taken to a sufficient depth in its progress under the river, to which the editor concurred: 'There is in our minds no question that want of depth has been the radical error in Mr Brunel's case, as in most others of the same kind.'[31]

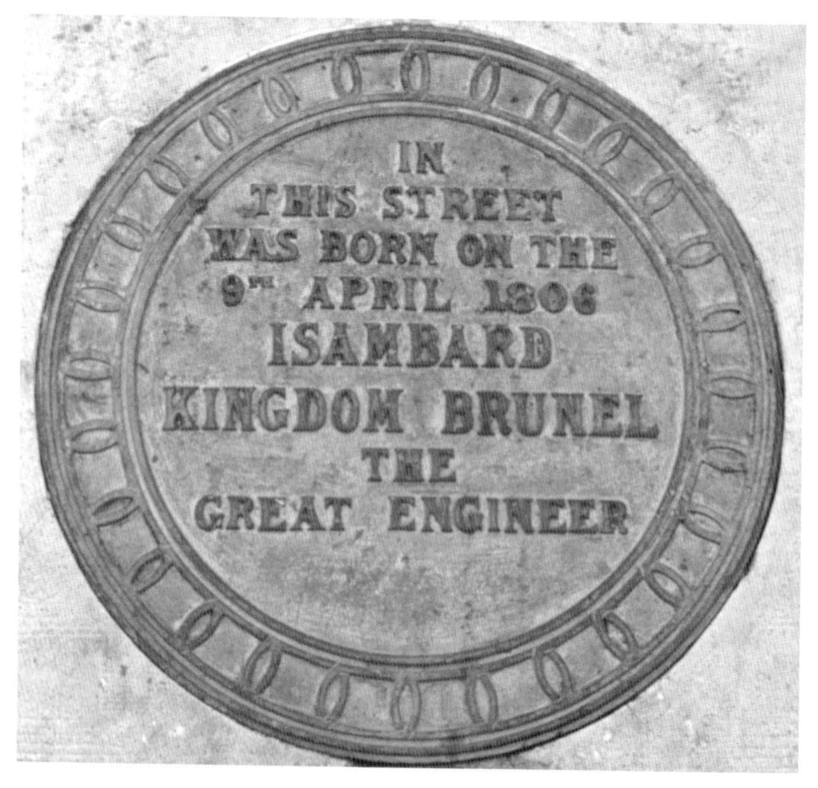

Plaque to Isambard Kingdom Brunel at Portsea, Portsmouth. (SKJ photograph)

A young engineer who only gave the initials 'S.Y.' wrote in to the *Mechanics Magazine* in January 1828, criticising the editor who had allowed remarks on the tunnel to be published, which he felt had been inconsiderate and illiberal. S.Y. went on to say,

> '... I must beg leave to say, that if errors in the estimate of any undertaking in the whole range of the profession of civil engineering can admit of excuse, it is certainly this of the Thames Tunnel. It is not like a bridge, or a canal, or any other work, of which sufficient have been executed to make the practical details familiar: it is the first work of its kind, in every respect – and the only one, to the best of my knowledge – which has ever been undertaken with sufficient method to render its success probable; and that method is entirely new. These things should be considered and allowed for.'[32]

The author goes on to comment on the 'errors' pointed out regarding the depth of the tunnel and the claims that the river bed was not properly examined. He adds that the neglect (of examining the river bed) has been sufficiently punished: 'Why, then, are other engineers to be called in? And what are they to be called in to do?'[33] He further comments that when mistakes are made in the construction of conventional works such as the building of a bridge, the public scarcely hear of it:

> 'But the engineer for the Tunnel is a man whose genius has hither to been exercised in a different branch of his profession: there is not fear of his bringing it to a successful termination, unless stopped for want of funds; and the plodders in the art feel mortified and degraded, that genius should accomplish an undertaking which has baffled the united efforts of experience and common-place understandings.'

Naturally the editor readily responded to this and addressed the question of 'united efforts':

> 'There have been but two other 'efforts' made, to which our correspondent can elude. The one was by Mr Dodd who, notoriously, was a man neither of 'common-place understanding' nor of 'experience', but a very ingenious schemer, without any practical talent whatever: the other was by Mr. Trevethick, [sic] whom nobody, that knows anything of him, would think of classing as inferior either in 'genius' or executive ability to Mr Brunel. It is well known besides, that what Mr Trevethick did, was merely to push a drift-way under the river, in order to ascertain whether a tunnel might be afterwards attempted with safety; and that as far as it went, the results were perfectly satisfactory.'[34]

Despite the 'advice' and solutions put forward, an interval of seven years was to elapse before work resumed and, in 1843, the tunnel was finally completed. Marc, recovering from the effects of a stroke, was able to attend the ceremonial opening on the 25 March 1843. The *Illustrated London News* referred to it as the 'eighth wonder of the world (if modern ingenuity had not extended the list ten fold)'. Marc's long

Portrait of Isambard Kingdom Brunel, engraving from a painting by John Callcott Horsley. Painted in 1838, it is one of the earliest portraits of Brunel, depicting him in his early thirties with plans of the GWR open in front of him. It is thus appropriate for his first commission in South Wales. (Courtesy of the National Railway Museum)

struggle had earned him a knighthood but the venture proved to be a financial failure. After twenty-two years as a pedestrian tunnel it was sold to the East London Railway and remains to this day a cross-Thames railway link. Marc passed away on 12 December 1849 at the age of eighty-one. While the Thames Tunnel is a major engineering achievement, the comment made by S.Y. of his genius being; '… exercised in a different branch of his profession' is true in as much that his pioneering work in the development of new manufacturing processes was to make the most significant contribution to Britain's industrial development.

Although the tunnel was his final work, his retirement was not spent in idleness. The wealth of his knowledge and advice was to be of the greatest help in the career of his son.

Isambard Kingdom Brunel had been born in the middle of a period of British history dominated by the Revolutionary and Napoleonic Wars, a period in which industry was to play an important role in expanding the economy and meeting the demands of war. This was to result in increasing mechanisation and a demand for mechanics and engineers. Marc's contribution in the field of mass production would not have been triggered without the necessity of war, while production at his factory, mass producing soldier's boots, came to an abrupt halt when peace was declared.`

Wartime demands and the ready availability of Government contracts provided the stimulus for mechanisation which was to permeate the whole shipbuilding sector and beyond. In the case of Marc's block mills it can be seen how this had laid the foundations of mass production.[35] Innovation crossover was evident in many areas with technological solutions being applied to problems, as in the example of John Wilkinson's cannon-lathe being applied to boring the cylinders required by James Watt's steam engine. This increased the supply and reliability of the steam engine which, in turn, greatly accelerated the introduction of steam power. As well as ships' blocks, the application of steam power for belt-driving machinery in the yards was to revolutionise a whole range of processes, such as victualling. Here the building of a ship's biscuit plant at the Deptford naval yard was to result in the Victualling Office of the Royal Navy becoming one of the biggest consumers of domestic agricultural produce.[36] The biscuit factory also became, like the block mills, one of the sights to be seen.[37] Another works was established close to Deptford in 1812 to manufacture iron chain-cable as a replacement for the hempen cables used by shipping. This purpose-built chainworks was developed by Capt. Sir Samuel Brown RN (1774-1851) at Millwall on the Thames. With regular naval contracts coming in, Brown had considered establishing a chainworks in every principal port but eventually settled on building his second purpose-built chainworks at Newbridge (later known as Pontypridd) to be closer to his main source of iron.[38]

Between 1788 and 1815, Britain's bar iron output rose from 32,000 tons to 150,000 tons, while pig iron output rocketed from just under 70,000 to 390,000.[39] The demand created by orders from the Board of Ordnance was to create enough work to keep ironworks, such as the Cyfarthfa ironworks of Merthyr Tydfil, operating at maximum output throughout the wartime period. The unprecedented opportunity for expansion presented in this period was to lead to the rise of Merthyr Tydfil as the leading industrial centre of its kind, making it well-placed to meet engineering demands for structural iron and the worldwide orders presented by railway construction from the 1830s onwards.

Preliminary Material and Chapter 1 Notes

1. Rolt, L.T.C., (1959). p.8 *Isambard Kingdom Brunel,* London: Longmans, Green & Co. '*Quand je serai grand, j'irai voir ce pays-là*' translates as 'when I grow up I want to go there for a visit.'
2. Gotch, Rosamund Brunel, ed. (1934). pp.168-69. *Mendelssohn and his Friends in Kensington, Letters from Fanny and Sophy Horsley written 1833-36,* London: Oxford University Press. Fanny Horsley writing in November 1834, '... and you may conceive John's delight in seeing Wales and the largest ironworks in the (country)'.
3. Royal Institution of Cornwall (RIC), Truro. Enys collection Trevithick papers. Trevithick to Giddy (Gilbert) 22 February 1804. Quoted (with corrected spelling) in: Trevithick, Richard, (1872). pp.161-62, (Vol.1), Life of Richard Trevithick with an Account of his Inventions. Vols 1 & 2, London: E & F.N. Spon.
4. Tomos, Dafydd, (1972). pp.25-26. *Michael Faraday in Wales including Faraday's Journal of his Tour through Wales in 1819,* Denbigh: Gwasg Gee. The chapter title comes from Alfred Tennyson's poem 'Locksley Hall', see Warburg, Jeremy. (1958). The Industrial Muse, p.23, London: Oxford University Press, 'When I went by the first train from Liverpool to Manchester (1830), I thought that the wheels ran in a groove. It was a black night and there was such a vast crowd around the train at the station that we could not see the wheels. Then I made this line.' Locksley Hall was published in 1842; the line was, 'Let the great world spin for ever down the ringing grooves of change.'
5. Rolt, L.T.C., (1959), p. *Isambard Kingdom Brunel,* Longmans, Green & Co., London.
6. Dowlais Iron Co. letterbooks, 1835 (1), letter 1109, 20 November 1835, I.K. Brunel to Henry Scale.
7. University of Bristol, Special Collections, IKB Letter Book. Rolt, L.T.C., (1959), *Isambard Kingdom Brunel,* Longmans, Green & Co., London.
8. Bessborough, Earl of, ed. (1950). Lady Charlotte Guest Extracts from her Journal 1833-1852. London: John Murray.
9. *Cardiff and the Marquesses of Bute,* John Davies, p.281.
10. *GWR magazine,* 1908, R. Price Williams.
11. 'When the Coal Comes From the Rhondda,' a traditional song dating from the Tonypandy lock-out of 1910-11 when 12,000 men from the Cambrian Combine went on strike. One version includes the following lines:
    'When the coal comes from the Rhondda
    Down that Taff Vale Railway line
    With my little pick and shovel
    I'll be there!
    I'll be there!'
12. Rolt, L.T.C., (1959), p.8. *Isambard Kingdom Brunel,* London: Longmans, Green & Co. '*Quand je serai grand, j'irai voir ce pays-là*' translates as 'when I grow up I want to go there for a visit.'
13. Rolt, L.T.C., (1959), p.8.
14. Noble, Celia Brunel, (1938), p.5. *The Brunels Father and Son,* London: Cobden-Sanderson.
15. Noble, Celia Brunel, (1938), p.12.
16. Rolt, L.T.C., (1959), p.10. Although Celia Brunel Noble refers to it as: '... the Park Theatre, with a cupola like the Corn Market in Paris.' Noble, Celia Brunel, (1938), p.17. Marc had submitted two designs for the Capitol building, *The Times,* 12 January 1984.
17. Noble, Celia Brunel, (1938), p.18. The letter in question was the letter 'S', no doubt Sophia's initial.
18. Noble, Celia Brunel, (1938), p.20.

19 Rolt, L.T.C., (1959). p.13.
20 Noble, Celia Brunel, (1938), p.24, quoted from *The Times*, 27 June 1814.
21 James, J.G., (1976), *Ralph Dodd, the Very Ingenious Schemer.* Trans. Newcomen Society.
22 Todd, A.C., (1967), p.89. *Beyond the Blaze; a Biography of Davies Gilbert.*
23 Lampe, David, (1963), p.19. *The Tunnel*, London: George G. Harrap.
24 Lampe, David, (1963), p.20. *The Tunnel*, London: George G. Harrap.
25 Trevithick, Francis, (1872), p.283. *Life of Richard Trevithick*, Vol.1, London: E. & F.N. Spon.
26 Buck, Alan, (1986), p.11. *The Little Giant: A Life of I.K. Brunel*, Newton Abbot: David & Charles.
27 Catalogue of Autograph Letters, (1996), p.16. Christie South Kensington, auction on 29 November 1996.
28 The Breguet Collections, (1998), p.28. Montres Breguet SA, L'Abbaye. The term 'chronograph' is derived from his work.
29 Trans. Newcomen Society, Vol.69, No.1, 1997-98, p.136.
30 Catalogue of Autograph Letters, (1996), p.37.
31 *Mechanics Magazine*, (1827-28), Vol.8, pp.204-05.
32 *Mechanics Magazine*, (1827-28), Vol. 8, pp.443-44.
33 *Mechanics Magazine,* (1827-28), Vol.8, pp.443-44.
34 *Mechanics Magazine*, (1827-28), Vol.8, pp.443-44.
35 Cossons, Neil, (2004), p.7. *From Manufacturer to Prosumer in 250 years.* Trans. Newcomen Society, Vol.74, No.1.
36 Duffy, M., (1980), p.78. *The Foundations of British Naval Power*, in Duffy (ed.), The Military Revolution and the State 1500-1800, Exeter: Exeter Studies in History, No.1.
37 Noble, Celia Brunel, (1938), p.24.
38 Jones, Stephen K, (1980). p.31. A *Link with the Past: the History of the Newbridge Works of Brown Lenox & Co., Pontypridd*, in Denning, Roy. (ed.), Glamorgan Historian. Vol.12. Barry: Stewart Williams.
39 Hyde, C.K. (1977). p.112-15. *Technological Change in the British Iron Industry*, 1700-1870. Princeton, Princeton University Press.

2

Merthyr: Iron Capital

'... The Largest Ironworks in the Country'[1]

While the industrialisation of Wales was confined, even by 1830, to a small number of centres or clusters of developments, Merthyr Tydfil, as one of those clusters, was to become one of the most important centres, as far as ironmaking was concerned, in this period anywhere in Great Britain and indeed the world.[2]

At its peak in the first half of the nineteenth century, Merthyr was the dominant force in the industrial profile of Wales. Even without the disruptive changes brought by the transition to steel, this profile was to change dramatically. Large-scale industrialisation became widespread in South Wales, with a growing demand for steam coal from the 1830s onwards. The year 1850 is seen as marking the decisive shift in the industrial base, from ironmaking to coal extraction, and the start of a second industrial era.[3]

From this point onwards, mining was to become the dominant industry in Wales, with the emphasis on export rather the local exploitation of a readily available and suitable fuel for steam-raising that could power manufacturing processes. The lack of mills, factories and workshops, with some prominent exceptions, such as the Brown Lenox chainworks at Newbridge, Pontypridd, reduced the opportunities to manufacture finished goods locally. Apart from rolled items such as rails, most of the iron production was to be refined as structural, engineering or finished products outside Wales.[4] Nevertheless, throughout the Industrial Revolution, Merthyr's position as a primary supplier of iron was without equal. Boasting the largest ironworks ever built, Merthyr was to become renowned as the ironmaking capital of the world.

Before this great period of expansion, Merthyr had a history and, indeed, a name derived from legend with the martyrdom of Tydfil, a Welsh princess who was the daughter of Brychan, King of Brycheiniog. Tydfil was murdered by foreign raiders around 1,500 years ago and the town of Merthyr Tydfil grew up around the shrine of the martyr Tydfil. Brychan was to give his name to Brecon which, by the eighteenth and nineteenth centuries, was a town that provided the majority of the banking and legal services needed by Merthyr's ironmasters.[5]

At its peak, the Iron Capital Merthyr Tydfil consisted of four separate ironworks known as Dowlais, Cyfarthfa, Penydarren and Plymouth. Ironmaking had centred

Cyfarthfa works in 1811. (From J.G. Wood's The Principal Rivers of Wales Illustrated, *London, 1813)*

on Merthyr Tydfil because of the abundance of iron ore, coal and limestone in the area. This mineral wealth was exploited by entrepreneurs in the latter half of the eighteenth century. Prominent among these were Thomas Lewis (1699-1764) of New House, Llanishen, founder of the Dowlais works, and Anthony Bacon (1718-86), who established the Cyfarthfa works. Of the promoters and ironmasters of the various Merthyr ironworks, Thomas Lewis was the first of the few native entrepreneurs. He was related to the well-known family of Lewis of Van, and had been associated with works at Caerphilly, Pentyrch and Melingriffith before founding Thomas Lewis & Co. at Dowlais in 1757. Lewis had a major advantage over other local adventurers in that he had sufficient capital to invest. The lack of local finance was one reason for the establishment of the three other ironworks by outside capitalists. Anthony Bacon hailed from Workington in Cumberland, another industrial area founded on its mineral wealth, and the four great families of ironmasters who followed Lewis and Bacon – Crawshay, Guest, Hill and Homfray – came to Merthyr from the Midlands and north of England as managers and promoters.

The Plymouth works had been started in 1763 by Isaac Wilkinson, John Guest and others who built a blast furnace on land belonging to the Earl of Plymouth. The works were taken over by Anthony Bacon in 1767. Richard Hill (*d.*1806) had worked for Bacon[6] at Cyfarthfa and, in 1784, had been installed by him as manager of the works at Plymouth. Under Hill and his son Anthony (1784-1862), the works attained a reputation for producing high-quality iron. Lack of capital prevented it from expanding into the major Merthyr concern, although additional furnaces were erected at Pentrebach and Dyffryn.

Merthyr with Cytarthfa Castle in the background, c.1830. (From South Wales Illustrated, *London, 1830)*

Much of the output of the Plymouth works was used by the early railways and by manufacturers such as Brown Lenox of Pontypridd. Using a special brand of wrought iron known as No.3 cable bolt iron, Brown Lenox produced chain cable for the Royal Navy and many merchant ships. Among these was Brunel's *Great Eastern* which required chain cable of what was a record size; 2⅝ in and 2⅞ in in diameter. Anthony Hill became a close friend of Isambard Kingdom Brunel through his connections with the Merthyr ironmasters, a friendship that was to continue for many years. In January 1858 Hill wrote to Brunel, congratulating him on the launch of the *Great Eastern* and urging him to take a well-earned break.[7]

In the first half of the nineteenth century, the Plymouth works were in the forefront of metallurgical progress under the scientific management of Anthony Hill. After Hill's death in 1862, the works were acquired by Messrs Fothergill, Hankey and Bateman, and survived until 1880, when they were finally overtaken by the steel age. The Plymouth works were put up for sale in 1882 but, with little interest being shown, it was decided to concentrate on exploiting the large reserves of steam coal on Plymouth land. Thomas Henry Bailey (1857-1939) was brought in to manage the collieries, with the enterprise becoming known as Hill's Plymouth Co. Ltd.[8]

In 1765 Anthony Bacon obtained land from Lord Talbot of Hensol and Michael Richards of Cardiff on very favourable terms – a rental of £100 p/a for ninety-nine years with no royalties. Surface leases held by local farmers were acquired by Bacon at rentals ranging from £100 to £200 p/a, in order to erect the buildings necessary for ironmaking. Bacon's enterprise became known as the Cyfarthfa works which, by 1803, was the largest in the world, employing 1,500 men.[9] The first furnace, known later as No.5, was built near outcrops of the lower coalfield seams, so it is likely that from the beginning this furnace used coke instead of charcoal. Pig iron, in great demand for the home market, was produced, and this helped to reduce imports from Sweden and Russia.

The iron was initially transported to Cardiff for shipment along the old road, and Bacon, together with other ironmasters, was instrumental in the turnpiking and improvement of this road under the Act of 1771. A late petition added this road to the Bill and, by the summer of 1775, work on improving the road, under the Llantrisant Trust, was well underway, with £500 having been expended, and Bacon taking two deed polls (at £100 each) to support this.[10] Improvements were carried out from Old Furnace (Tongwynlais), where it joined the Cardiff district's turnpike road, to Merthyr Tydfil. The main line of communications in terms of heavy traffic was naturally north to south to enable the ironmasters to despatch their iron through the shipping places at Cardiff, but there was also traffic across the hills in semi-finished materials from the smaller ironworks at the heads of the Monmouthshire valleys. Merthyr also tended to look north to Brecon for its banking and legal services rather than to Cardiff at this period. The start of the American War of Independence in 1775 created a demand for ordnance and a boring mill was set up by Bacon. In 1782, however, Bacon was forbidden to manufacture ordnance, when an act preventing Members of Parliament from holding Government contracts was passed (Bacon represented Aylesbury from 1764). Bacon circumvented this by leasing the mill to Francis Homfray (1726-98), an ironmaster from Worcester. The relationship between Homfray and Bacon was difficult, however, with the former leaving Cyfarthfa to support his two sons, Jeremiah and Samuel, in establishing a works at Penydarren in 1784. The lease of the mill was transferred to David Tanner of the Tintern Abbey Wireworks, who, like Homfray, stayed less than two years.

Richard Crawshay (1739-1810) was taken into partnership at Cyfarthfa by Bacon around the period 1777-80 and, following the death of Anthony Bacon, was to run the works for Bacon's sons. He obtained the lease of the whole undertaking from Bacon's sons and become sole proprietor of the Cyfarthfa works in 1794. Crawshay was a Yorkshireman who fell out with his parents and, at the age of sixteen, left for London, where he started at the bottom of the iron trade, working in a warehouse selling flat irons. In the Samuel Smiles tradition he worked his way up to become the owner of the warehouse and became a wealthy man. Having money to invest brought him to Merthyr, while he used his trade connections to establish a selling agency for Cyfarthfa iron at George Yard, Upper Thames Street, London. Crawshay introduced processes for puddling and rolling iron, based on the pioneering work of Henry Cort, and new furnaces and forges were erected at Cyfarthfa. The demand for cannon and manufactured iron goods continued throughout the Napoleonic wars. Richard Crawshay's son and grandsons carried on building up the works after his death, particularly the enterprising William Crawshay II (1788-1867).

William Crawshay II has passed into history as the 'iron king' of Merthyr, building the castellated mansion he called Cyfarthfa Castle in 1825 at the height of the prosperity of the works. Today, the original part of Cyfarthfa Castle houses a museum and art gallery from which one can look out across the valley towards the monolithic remains of the Cyfarthfa ironworks that stand as a monument to the Crawshays and their empire. As well as the large furnace bank, a cast-iron bridge (Pont-y-cafnau) can also be found crossing the Taff. This bridge not only carried a tramroad bringing limestone into the Cyfarthfa works, but also water as part of the feeder system to the works. The Crawshays had the widest empire of the Merthyr ironmasters, for, apart from the Cyfarthfa works (including a subsidiary works at Ynysfach in Merthyr) and the London House, they had ironmaking interests at Hirwaun, Treforest and the Forest of Dean, with another branch of the family engaged in manufacturing at Gateshead.

The ironworks at Cyfarthfa continued under William Crawshay II's son, Robert Thompson Crawshay (1817-79), until 1874, when the works closed due to trade union agitation, and only reopened after R.T. Crawshay's death. The works had not changed to steel production due to his unwillingness to abandon ironmaking, and it was left to his sons to attempt the conversion to steel production. In 1902 the company was acquired by Guest, Keen and Nettlefold (of the Dowlais works), and closed in 1910, only to briefly reopen in 1916 to make war materials. The other outposts of the Crawshay empire were also to fail with the coming of the steel age.

As a replacement for wrought iron in the manufacture of rails, steel had been making considerable inroads since the late 1850s, and industry was increasingly demanding steel for engineering purposes. The construction of the Forth Railway Bridge was to mark the end of iron's dominance and, by the 1890s, very little wrought iron was being rolled for structural purposes. Iron had always been the Crawshay's metier, and that trade was fast drawing to a close.

In Merthyr, the first ironworks to close was Penydarren in 1859, a works founded by Jeremiah and Samuel Homfray in 1784. Their father, Francis Homfray, had come from Wollaston Hall in Worcestershire to lease the Cyfarthfa cannon-boring mill. He then went on to help his sons acquire a lease of land at Penydarren which was rich in iron ore deposits. His two sons, Jeremiah (1759-1833) and Samuel (1765-1822), built up the works into a volume producer despite the lack of its own coal resources and water power. The adjoining Dowlais works had first call on the brook that ran through both works, resulting in constant agitation between the two companies over issues such as the water supply and mineral leases, which was to lead to frequent lawsuits, most of which were lost by the Homfrays.

Francis Homfray was to play an important part in starting the next phase of transport improvements, being given the credit for suggesting the building of the Glamorganshire Canal, although Crawshay was to be the main driver of this enterprise.[11] It is likely that Homfray knew Thomas Dadford, the canal engineer, from his previous enterprises in the West Midlands, and recommended him to Richard Crawshay in 1788. Within two years of taking over the lease of Cyfarthfa, Crawshay was considering the best means of improving the mode of transport between Merthyr and Cardiff. He wished to avoid the excessive costs of transporting his iron by road and felt he was being held to ransom by hauliers such as William Key.[12] Crawshay persuaded the other ironmasters to support his scheme and, although all were not completely enthusiastic, they were to contribute to the cost of Dadford's survey.

In 1790, Crawshay's Bill for the construction of the canal from Merthyr to Cardiff was passed by Parliament, and by 1792 the canal was open to Treforest. Two years later it was completed to Cardiff, although direct access to the sea was not possible until 1798 when the sea-lock extension was opened. The construction was undertaken by Dadford, with his son, Thomas Jr, and Thomas Sheasby as joint contractors. Due to the severe gradients presented by the terrain between Merthyr and Cardiff, over fifty locks had to be constructed, sixteen of which were used between Quakers Yard and Abercynon, to overcome a drop in level of 200ft. As a region, South Wales was now a canal economy: the major emphasis was that: '... the canal-based economies became more specialised, more differentiated from each other and more internally unified.'[13]

Samuel Homfray was to become the sole manager at Penydarren, while his brother started the Ebbw Vale works. Despite their enthusiasm and enterprise as ironmasters (Samuel was also involved in the establishment of the Tredegar ironworks), quarrels and legal disputes, even between themselves, were frequent. Samuel developed a method of making 'finer's metal' in 1794, a principal stage in the manufacture of bar iron, and he worked continuously to introduce the latest technology at Penydarren and Tredegar ironworks. He was to overcome one of the major sources of dispute, over adequate coal supplies, by marrying the daughter of Sir Charles Gould Morgan of Tredegar Park and obtaining a lease of 3,000 acres of mineral land very cheaply.

After Samuel Homfray's death in 1822, the works were continued by the other partners, William Forman and Alderman William Thompson. In 1830 rails were rolled for the Liverpool and Manchester Railway, and in 1839 the six furnaces in blast sent a total of 15,540 tons of iron down the Glamorganshire Canal. This can be compared with a combined total of over 100,000 tons transported on the waterway in the same year by the nine blast furnaces at Cyfarthfa and Ynysfach, the seven at Plymouth and Dyffryn, and the fourteen at Dowlais. In 1859 the Penydarren works were acquired by the Dowlais Iron Co., and closed down.

Church memorial to Sir John Guest in Dowlais Parish church.

Dowlais House, Merthyr, around the mid-1870s. (Courtesy of Glamorgan Record Office)

The latter company was the oldest and ultimately the most successful of the iron-making concerns in Merthyr. Following Thomas Lewis's mineral lease of 1757, articles of co-partnership were signed on 19 September 1759, by nine partners, Thomas Lewis being the major investor with two other South Wales men, Thomas Price of Watford (Caerphilly) and Richard Jenkins of Cardiff. There were also four Bristol partners and another partner who was one of the famous names of iron-making, Isaac Wilkinson. Wilkinson allowed the use of his patented designs for blowing the furnaces at Dowlais in return for a payment if the iron produced each week exceeded 20 tons. Prophetically, Wilkinson had once remarked in 1779 to a young friend that '… you will live to see waggons drawn by steam. I would have made such a waggon myself if I had had time.'[14] The first small furnace at Dowlais was producing 18 tons of iron a week by the end of 1760. In 1763 a new mineral lease was taken out from Lady Windsor by the Dowlais Co. This also allowed 22 acres of common land to be built upon for ironworks.

In 1767, John Guest (1722-87) became manager of the works, and then a partner in 1782. Guest, originally from Broseley in Shropshire, was the first of the family to be associated with the works. Charles Wood was the resident managing agent for the Cyfarthfa works from 1766-74, and he records the following in his diary on the 6 May 1767:

> 'Mr. John Guest & his Brother Thomas were here this day the former informed me that he had taken the management of Dowlais furnace for the term of 14 years, but he did not inform me of the Terms only that he was to deliver the Pigs in the bank at a certain price & to be paid Monthly for them. And he said he would be glad to be upon good Terms with me & do all we each of us could for the benefit of both which I told him was my desire.'[15]

Guest and his followers appear to have raised their profile as 'foreigners' in the locality as another diarist, William Thomas, was to remark in August 1767 when he referred to a brickmaker then working on Coedriglan House:

> 'This man that burns is an Englishman from Shropshire and married to a widow at Mirthyr Tydfil [sic]. He came to Mirthyr Tydvil with other Englishmen these few years past to make there the New furnaces etc., being that place is swarm by Englishmen since the Iron work came there.' [16]

Celebrating the bicentenary of Dowlais works in 1959, *The Dowlais Story* refers to John Guest as a: '... robust, hard-working man. Relatives and friends soon joined him at Dowlais, taking on jobs as moulders, refiners, clerks. It began a family tradition which has survived the centuries.'[17] Two of his younger brothers followed John Guest to Merthyr and in the following years their sons were to hold key positions in the works. It was seen as a sensible approach, not only by Guest but by other ironmasters, in relying on family loyalties to prevent fraud and embezzlement or as William Taitt (1748-1815), the son-in-law of John Guest, put it: 'so that we may not be plunder'd'.[18] The Guest name continues to this day as part of the title of Guest, Keen & Nettlefolds, a major engineering company. Before it was assigned to Anthony Bacon, Guest had been associated with Isaac Wilkinson in the original lease for the land owned by the Earl of Plymouth. He began to build up the works and make technological advances, such as the adoption of a process from his brother-in-law, Peter Onions, for converting cast into wrought iron in 1783. He and his son Thomas Guest (1749-1807) increased their holdings in the works. Following his

Remains of Dowlais House with the Guest Memorial Library in background, photographed in 1977. (SKJ photograph)

Dowlais blast furnace. (Courtesy of Glamorgan Record Office)

father's death, Thomas became joint manager with William Taitt. Taitt was a vigorous driving force at Dowlais and continued the expansion started by John Guest. During their joint management, a steam engine for providing a blast to the furnaces was ordered from Boulton & Watt in 1798.

Taitt's death in 1815 and the transference of his shares to Thomas Guest's son, Josiah John Guest (1785-1852), left only the descendants of Thomas Lewis and the Guest family as partners in the Dowlais works. Josiah John Guest was born in Dowlais and he assumed his father's position as joint manager in 1807. Guest raised the works from third place behind Penydarren and Cyfarthfa to first place as the largest iron producer in the world. One order in 1821 was for rails for the Stockton and Darlington Railway, and Guest was to become involved in the building of the first modern railway in Wales, the Taff Vale Railway (the story will be examined in greater detail from chapter 5 onwards). While the line engineered by Isambard Kingdom Brunel was the only standard-gauge (4ft 8½in) railway he constructed in this country, his broad gauge (7ft ¼in) was to reach Merthyr in 1853 in the shape of the Vale of Neath Railway.

Guest became the sole owner of the Dowlais works in 1851, but died the following year. Following his death, his wife, Lady Charlotte Guest (1812-95), took control of the works. She directed the works until the spring of 1855 when responsibility was passed to trustees, the resident trustee being George Thomas Clark (1809-98). Clark was an engineer who had worked under Brunel on the GWR and he, together with the new manager, William Menelaus (1818-82), brought the works into the steelmaking era.

Dowlais was the first ironworks to adopt the process of converting iron into steel, pioneered by Sir Henry Bessemer (1813-98), and produced its first steel ingot in 1865. This was used for rolling steel rails, but to make steel by this process required haematite iron ores in quantities that could not be obtained locally. The need to import iron ore from the Dowlais Iron Co.'s own mines in Spain forced the company to look seriously at the future of Merthyr as a steelmaking centre. Eventually it was decided to build a new steelworks, not at Merthyr but alongside the docks at Cardiff, where raw materials could be unloaded directly into the works. This was one of the last major changes seen through by G.T. Clark, and the steelworks, known officially as 'Dowlais by the Sea', but more commonly as 'East Moors', opened in 1891.

The 1920s and 1930s saw wholesale reductions in employment in South Wales; in the coal industry alone the number of men employed fell from 250,000 to just over 100,000. The iron industry was also going through a major upheaval and, in 1930, all iron and steel production ceased at Dowlais, and only the foundry and fitting shops remained in work six years later. At this time the closures had brought widespread levels of unemployment and deprivation to the town. Indeed, on a visit in 1936 the uncrowned Edward VIII is reported to have uttered the words: 'something must be done' as he viewed the desolation in the town. The King's comments and increased political pressure led to the suggestion of a radical plan by a Permanent Secretary, to move the entire population of Merthyr to a new town on the river Usk or on the Glamorgan coast. The plan was taken so far as to be costed at £10 million but it was eventually decided that intervention would be the Government's policy to help the people of Merthyr without moving them out.[19] Cardiff's East Moors steel works was to close at the end in 1978. The closure in December 1987 of the Dowlais casting foundry, the last working entity of the iron capital, Merthyr Tydfil, brought 228 years of continuous production to an end.[20]

When John Callcott Horsley (1817-1903) was being taken on trips to South Wales by Brunel, his sister referred to Merthyr as '… the greatest ironworks in the country', which was an understatement as the Dowlais works alone was the greatest ironworks in the world. In 1846 it had a workforce of nearly 7,000, an annual wage bill of over £250,000 and operated eighteen blast furnaces[21]. This position was achieved through steady progress and investment from the beginning of the nineteenth century. It started as a modest supplier of pig iron to the trade with some output in cast-iron ware and iron rails, and grew to its dominant position when it overtook Cyfarthfa in the 1830s.

It was with its production of iron rails that Dowlais was to see the foundation of its future prosperity. The all-iron rail, in this case the cast-iron edge rail, was a South Wales innovation, even if it was based on the wooden, cast-iron plated track first used at Coalbrookdale. These rails were used on the Dowlais works feeder-line that connected with the Glamorganshire Canal. In 1791 William Taitt described the rails used:

'We are now making Rails for our own Waggon Way, which weigh 44 li or 45 li per yard. The Rails are 6 feet long, 3 pin holes in them, mitred at the ends, 3 Inches broad at Bottom, 2½ Inches top, and near 2 Inches thick. If this sort will do (they are cast in flask) we will deliver them in London for £8. 10. 0d per ton & would send some of them immediately.'[22]

This type of rail was designed for flanged wheels and was the forerunner of the modern rail. It was adopted in a number of cases for the numerous canal feeders being built to directly connect with the canals that were being constructed at the end of the eighteenth century. The Monmouthshire Canal also had a considerable mileage of connecting railways laid with this type of rail.[23] The alternative was the 'L'-shaped rail or plate rail designed for unflanged wheels. Benjamin Outram of the Butterley ironworks was the main advocate of this kind of rail and was to build his railways using this kind of rail, and recommend it to anyone who sought his advice. His advice was sought on the Monmouthshire Canal railways in 1799 and this led to a changeover from edge rail to plate rail. Almost all the rest of South Wales and further afield followed suit and, by the 1820s, the plateway was dominant, except in the north-east of England.

Although a major transport improvement had been the building of the Glamorganshire Canal, one section of the canal in particular was to become a bottleneck for traffic and the cause of frequent disputes with the Crawshays over their attempts to control the running of the canal for their benefit. This eventually forced the other ironmasters to build an alternative, in this case a tramroad. Trouble had been brewing for some time; Richard Hill complained about the canal company taking water from the Taff that was legally his; Dowlais frequently complained about Cyfarthfa boats taking preference to theirs, and so it went on. In June 1798 matters were to come to a head at the Annual General Meeting of the Glamorganshire Canal, when the leading dissidents, representing the Dowlais, Penydarren and Plymouth ironworks, failed to be re-elected and were replaced. The dissidents took action and, on 24 September 1798, a parliamentary notice appeared for leave to make a tramroad.[24] This was to virtually parallel the canal from Merthyr to Cardiff, and have branches coming off at Quakers Yard to Carno Mill at the head of the Rhymney Valley, and to Hirwaun in the Cynon Valley. Naturally, the canal company moved quickly to oppose this rival and they joined forces with the commissioners of the Cardiff to Merthyr turnpike to issue a joint statement appealing to local landowners. Charles Wilkins repeats part of this statement which concludes with:

'Whoever looks candidly on this attempt will see the injury done to two useful sets of men, viz., the Canal and Turnpike Road subscribers. A waste of lands to profit nobody! Whatever may be urg'd per contra, we are of opinion that by a road the iron will never will be carried so cheap as it now is by the canal, and that in three years time it will be considerably lowered by the increase of trade upon it.' [25]

The Bill was withdrawn before the second reading but, even while the Bill was making its initial progress through Parliament, the ironmasters had decided to submit a pared-down plan in which the Hirwaun branch would be dropped:

> 'Plan of an Intended DRAM ROAD from or near Carno Mill in the Parish of Browellty [sic] & County of Monmouth to or near the Sea Lock below the Town of Cardiff with a Branch from the same to the Limestone Rocks in the Parish of Merthyr Tydvill & the County of Glamorgan. By P. Williams 1799.'[26]

It was to be reduced even further by the dropping of the Carno Mill leg and, finally, to a section of tramroad running just between Merthyr and Navigation (Abercynon). This is what was eventually built and was to be known as the Penydarren or, more correctly, the Merthyr Tramroad. In March 1799 a Bill sponsored by the partners of the three ironworks (Plymouth, Penydarren and Dowlais) was withdrawn due to the combined opposition of both the Glamorganshire and Monmouthshire Canals. The tramroad, however, was constructed without an Act, which could legally be

A flight of locks looking down to Abercynon. Although this photograph is from October 1913 when the canal had been closed for some fifteen years, the main lock staircase, a group of eleven locks in ¼ mile, (and a major bottleneck to traffic) can be clearly shown. In total, the sixteen locks at Abercynon raised the canal some 200ft in the distance of a mile. The Merthyr Tramroad was to bypass the canal at this point and, on the left, can be seen the main incline of the TVR, Brunel's answer to overcoming the difference in level. (Photograph by W. Rowlands, SKJ collection)

Canal boats carrying patent fuel in the West Bute Dock. (Photograph by H.J.B. Wills, SKJ/SR collection)

done under the 4-mile clause of the Glamorganshire Canal's Act. Most of the land needed was owned by Lord Plymouth who did not oppose its construction and there was no need to invoke compulsory powers to construct the tramroad. It ran from a junction with the existing Dowlais Tramroad at Merthyr to the canal basin at Abercynon. George Overton was the engineer of the 9½-mile-long tramroad that was to be built by Richard Hill under the direction of William Taitt of Dowlais, and which opened in 1802 as a 4ft 2in-gauge plateway. As part of the wholesale rejection of the edge rail in South Wales that had occurred a few years previously, the line of the Merthyr Tramroad was to consist of cast-iron L-shaped plates which were laid to the gauge of 4ft 2in, measured inside the plate flanges or 4ft 4in over them.[27]

However, up to this time, the railway, as edge rail or plateway, was used primarily as a feeder to canals. In 1804 an event took place in Merthyr that was to show the real potential and future of the railway. Naturally, it was an event in which the Merthyr ironworks were involved, although the role of the Dowlais works was largely a background one based on the details already outlined. Two of the other three ironworks, led by their ironmasters, would be involved on the fringe but the direct role, one described as showing, '… spirited and active patronage'[28] in bringing forward an innovation of worldwide significance, was by the Penydarren ironmaster, Samuel Homfray.

CHAPTER 2 NOTES

1 Gotch, Rosamund Brunel, ed. (1934), pp.168-69. *Mendelssohn and his Friends in Kensington, Letters from Fanny and Sophy Horsley* written 1833-36. London: Oxford University Press. Fanny Horsley writing in November 1834, '... and you may conceive John's delight in seeing Wales and the largest ironworks in the (country)'.

2 Williams, John, (1995). pp.14-35. *Was Wales Industrialised.* Llandysul: Gomer Press. This chapter is opened by referring to the assertion made by the historian Sir Frederick Rees that industrialisation in Wales was still confined to a few, localised pockets by 1830, a date that marks the end of the classic Industrial Revolution period. Rees, J.F., (1947). Studies in Welsh History, Chapter IX, Cardiff.

3 Davies, John, (1996), p.116. *The Making of Wales.* Stroud: CADW/Sutton Publishing Ltd.

4 Cannons were sent to be bored elsewhere.

5 Lloyd, John, (1906). *The Early History of the Old South Wales Ironworks (1760-1840),* London: Bedford Press. John Lloyd compiled this history from the records of one firm of Brecon solicitors: Messrs Walter and John Powell. These papers are now preserved as the Maybery papers.

6 Richard Hill (*d.*1806) had married the sister of Bacon's mistress. See Rowson, Stephen & Wright, Ian L., (2001), p.71. *The Glamorganshire and Aberdare Canals.* Vol.1, Lydney: Black Dwarf Publications.

7 Buchanan, R.A., (2002), pp.205-06. *Brunel, The Life and Times of Isambard Kingdom Brunel,* London: Hambledon and London.

8 Perkins, John, Thomas, Clive and Evans, Jack (1986), p.123. *The Historic Taf Valleys,* Vol.3, Thomas, Clive. *Social and Industrial History,* Bridgend: Merthyr Tydfil and District Naturalists' Society.

9 Cyfarthfa is Welsh for 'barking'. The area where the works was set up was known as the 'barking place of dogs'. Taylor, Margaret Stewart, (1967), p.13. *The Crawshays of Cyfarthfa Castle: A Family History,* London: Robert Hale.

10 John, A.H. & Williams, Glanmor, (1980), *Glamorgan County History, Vol.V, Industrial Glamorgan,* Pollins, Harold, p.427, *Chapter IX, The Development of Transport 1750-1914.*

11 Lewis, Samuel, (1840), *A Topographical History of Wales.*

12 Rowson, Stephen and Wright, Ian. L., (2001), p.17, *The Glamorganshire and Aberdare Canal,* Vol.1, Lydney: Black Dwarf Publications.

13 Berg, Maxine, (1994), p.101, *The Age of Manfactures 1700-1820,* London: Routledge.

14 Lewis, M.J.T., (1970). p.296. *Early Wooden Railways,* London: Routledge & Kegan Paul.

15 Gross, Joseph, ed. (2001). p.190. *The Diary of Charles Wood of Cyfarthfa Ironworks, Merthyr Tydfil, 1766-1767,* Cardiff: Merton Priory Press.

16 Denning, R.T.W., ed. (1995), p.192, *The Diary of William Thomas (of Michaelston-super-Ely) 1762-1795,* Entry for 2 August 1767, Cardiff: South Wales Record Society and South Glamorgan County Council Libraries & Arts Department.

17 Woodhouse, N., (1959), *The Dowlais Story 1759-1959,* Cardiff: Western Mail & Echo.

18 Evans, Chris, (1993). p.62-3. *The Labyrinth of Flames: Work and Social Conflict in Early Industrial Merthyr Tydfil,* Cardiff: University of Wales Press.

19 Rowlands, Ted, (2000), *Something **must** be done: South Wales v Whitehall 1921-1951,* Merthyr Tydfil: TTC Books.

20 John A. Owen was the last manager of what had become a general castings foundry, converted in 1971 to making castings in liquid sand. Owen, John A. (1972, reprinted 2001), *A Short History of the Dowlais Ironworks,* Merthyr: Merthyr Tydfil Library Service.

21 Carr, J.C. and Taplin, W., (1962), p.8, *History of the British Steel Industry,* Oxford: Basil Blackwell.

22 Glamorgan Record Office (GRO), Dowlais letterbook, copy letters 1782-94, p.315, William Taitt to William Hawks, 17 March 1791.
23 Lewis, M.J.T., (1970), p.293.
24 Hadfield, Charles, (1967), p.97. *The Canals of South Wales and the Border*, Newton Abbot: David & Charles.
25 Wilkins, Charles, (1888), p.183. *The South Wales Coal Trade and its Allied Industries*. Cardiff: Daniel Owen.
26 Institution of Civil Engineers library archives, London.
27 Rattenbury, Gordon and Lewis, M.J.T., (2004), p.48, *Merthyr Tydfil Tramroads and their Locomotives*, Lewis, M.J.T. The Locomotives. (Book II). Oxford: Railway and Canal Historical Society. The revised and expanded Book II of this publication was originally published as *Steam on the Penydarren* (April 1975), Industrial Railway Record, No.59.
28 *The Cambrian*, 25 February 1804.

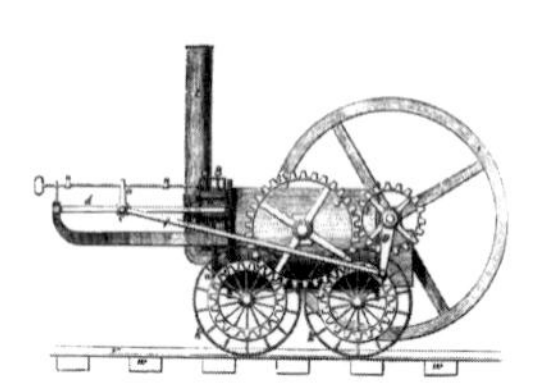

3

TREVITHICK'S DRAGON

'-THE PUBLICK UNTILL NOW CALLD MEE A SCHEMEING FELLOW... '[1]

The trials of Richard Trevithick's locomotive from 13 February 1804, and the historic run on 21 February 1804, marked the date of an innovation that would have enormous significance – the first steam locomotive to run on rails anywhere in the world. By running and hauling a train of loaded wagons on the Merthyr Tramroad from the Penydarren ironworks in Merthyr, to the canal basin at Abercynon, it had completed the first authenticated steam railway journey. There were technical problems with the locomotive engine itself, which was to prove too heavy for the cast-iron tram plates it ran on, but it clearly demonstrated that Trevithick's high-pressure steam engine was the key to the development of steam locomotion and that a smooth wheel running on smooth rails was a practical proposition. Considerable problems still had to be overcome to make steam locomotion a commercial proposition, an objective that Trevithick pursued at Penydarren and elsewhere, but success was to elude him. The initiative to solve these problems and produce a commercial solution passed to others such as Blenkinsop and the Stephensons. The way had been shown, however, and these later pioneers were to acknowledge their debt to Trevithick.

In many ways the seeds of this innovation can be seen to have resulted from the need to improve the transport infrastructure of industrial South Wales, in particular those serving the ironworks of Merthyr Tydfil. In the same year that the Merthyr Tramroad opened (1802), Richard Trevithick and Andrew Vivian took out a joint patent for 'Methods For Improving The Construction of Steam Engines, And The Application Thereof For Driving Carriages And For Other Purposes.'[2]

A major innovation was the simplicity of the high-pressure engine, '... what could be said when Trevithick dispensed with the beam, air pump, acted directly on the crank-shaft, increased the speed, decreased the space needed, made it capable of being manufactured in small powers and of going anywhere, doing anything cheaply?'[3] There was the small matter of safety; one of Trevithick's engines, which had been installed to drive a pump at Greenwich, exploded in September 1803, killing four workmen. It was found that the engine boy had fixed the safety valve down while he went off to fish for eels.[4] The risk of explosion was seized on by Boulton & Watt who, according to Trevithick,

> '... had strained every nerve to get a Bill in the House to stop these engines, saying the lives of the public are endangered by them, and I have no doubt that they would have carried their point if Mr. Homfray had not gone to London to prevent it...'[5]

Writing almost thirty years later, Elijah Galloway was to note:

> 'The extended use of such engines at the present day [1831] proves that the public opinion is materially changed in regard to their utility and safety. The risk of explosion was a drawback upon them...'[6]

Galloway goes on to explain the benefits of the high-pressure steam:

> 'Water does not boil under a temperature of 212° of Fahrenheit, at which temperature its force, when confined, is barely equal to that of the atmosphere. But let the temperature be increased only 38° more, (which can be effected with a comparatively small addition of fuel,) and its force will be 28lbs. to the square inch. In like manner let the temperature be increased to 280°, and the force will be equal to 56lbs. Thus, the increased force far exceeds the increased consumption of fuel, and, consequently, the greater the pressure of the steam, the greater will be the saving'.[7]

The reasons for concern about safety and the risk of explosion can be imagined: some idea of the boiler pressure above can be related to that of the tyre pressure of a modern car, which would be around 33lbs to the square inch! It was no wonder that Trevithick and Vivian referred to it as 'strong steam' and took special care in the design of the boiler, '...made of a round figure, to bear the expansive action of strong steam.'[8] It was also an expression used by Davies Gilbert who was a fellow Cornishman and mentor to Trevithick. Due to his appreciation of scientific matters and his role as a Member of Parliament, Gilbert was to contribute greatly to Britain's role as an industrial nation. As well as Trevithick, he had supported and mentored other Cornish engineers such as Jonathan Hornblower (1752-1815). Gilbert had more than once urged upon Hornblower, who was busy working on his rotary steam engine, that he would have more success if he used the steam when it was expanding. Trevithick, however, picked up the advice given and '... within a Month several "Puffers" were in actual work.'[9] 'Puffers' was Gilbert's description of the sound made by the high-pressure engines at work which were first used to raise ore up from the mines.

The compact size of the high-pressure engine compared to the power it could deliver, led Trevithick to take a momentous leap from stationary engine to locomotive and, towards the end of 1800, he began to construct his first 'travelling engine' at Hayle Foundry. The famous journey of this road vehicle or 'Dragon' (another Gilbert description) was to take place over Christmas 1801 on Beacon Hill, Camborne. It set out on the 28 December and although it never reached its destination, breaking down and catching fire on the way, it had demonstrated that steam locomotion was a reality. Before another 'Dragon' was constructed, Gilbert urged Trevithick to patent his ideas, which he did with his cousin Andrew Vivian.

Trevithick was to receive several orders for his high-pressure engine, which was to be used for milling corn and boring brass cannons, while his second road 'dragon' ran through the streets of London. With such heightened publicity, there was great interest in the potential of the engine and, in 1803, Samuel Homfray engaged him to build a high-pressure steam engine at his Penydarren works. If satisfied, he would take a share in Trevithick's patent. The support shown by Samuel Homfray was recognised by *The Cambrian* newspaper the following year in reporting on one of Trevithick's engines in London; '... we have great pleasure in adding, that this valuable improvement has attained its present perfection in consequence of the spirited and active patronage of Mr Homfray of Merthyr Tydfil, Glamorgan'.[10] Homfray was keen to develop new technologies and had even built steam pumping engines, one of which he supplied to the Tarnowitz lead mine in Germany in 1786.[11] The ironmaster recognised the opportunities that the high-pressure boiler presented, including the possibility that it could be fitted to run on rails, and Trevithick had already demonstrated twice that it could be run as a road vehicle. On 1 October 1803, Trevithick wrote to Gilbert:

> 'There will be a realroad engine at work here in a fortnight, it will go on reals not exceeding an elevation of one-fiftieth part of a perpendicular and of considerable length. The cylinder is 8½ in diameter, to go about two and a half miles an hour, it is to have same velocity of the piston rod. It will weigh, water and all complete, within 5 tons.'[12]

Trevithick urged Gilbert to come to Merthyr to witness at first-hand the performance of his rail 'Dragon', tempting him to leave Oxford: '... I dought not but you will be fully satisfyde for the toil of the journey by a sight of the engine...'[13] This was written the day before the historic journey on 21 February 1804. Trevithick finally got Gilbert to come down when he appealed for his support in dealing with the expected arrival of Government engineers who wished to inspect the engine. With the offer of a well-aired bed and a hearty welcome from Homfray, Gilbert arrived on 12 March and, while he concluded that '... the engine worked with success...' [14], the extensive damage caused by the weight of the engine to the cast-iron tram plates was to create considerable concern as to the way forward. Over what length of run and under what load Gilbert was able to see the engine is difficult to determine; it was probably no more than 2 miles as in the case of the first trials.

Gilbert was, however, impressed by the system of tramroads employed by Homfray at Penydarren and Tredegar and, indeed, throughout the ironworks of Merthyr. He discussed the subject with Homfray and was to seek details from his manager, William Wood, then and in the future on how to construct a 'railroad'. The result was that some years later the first tramroad in Cornwall was laid between Poldice Mine, near Redruth, and Portreath Harbour.[15]

This takes us slightly ahead of events and how the story unfolded locally. When talk of the possibility of a railroad engine reached the ears of the other Merthyr ironmasters, great debate took place on the practicability of such a revolutionary scheme and wagers were made. This part of the story, which according to legend provided the initiative for the historic run, has several versions. The variations concern the individuals who placed the bet, but the most likely scenario is that set out by Stanley Mercer;

> 'History has it that Samuel Homfray bet Richard Crawshay 500 guineas that he would take a load of iron to the basin by means of an engine instead of a horse. Crawshay bet a similar sum that it could not be done and Richard Hill held the stakes which thus amounted to 1,000 guineas. After the journey was made, Hill was not at first agreeable to make the award, asserting that as Trevithick had moved the tram plates in the Plymouth tunnel to bring the engine to the centre of the arch, the line was not in the same position as when the bet was made.' [16]

Charles Wilkins, one of Merthyr's most prolific historians, covers this famous episode in Merthyr's history in all his books, such as *The South Wales Coal Trade* and *The History of Merthyr*, but he refers to a bet of £1,000 made between Samuel Homfray and Richard Crawshay.[17] Trevithick himself refers in a letter to: 'The gentleman that bet 500 guineas against it… '.[18] Wilkins probably takes his figure from one of the eye-witness accounts to the event, as Rees Jones quoted that same figure when interviewed in 1858.[19] Wilkins adds to the role played by Trevithick, '… who, according to the terms of the wager, had to control and repair the strange monster unaided...'

Richard Crawshay (1739-1810) is sometimes left out as the second party to the wager. L.T.C. Rolt, in his biography of Trevithick, *The Cornish Giant*[20], states that the bet took place between Homfray and Hill, possibly because the second party to the wager is not mentioned by Trevithick. It is clear from the letters written by Trevithick that 'Mr Hill' acted as the stakeholder, but frustratingly he does not mention the second party by name: '… the gentleman that bet 500 guineas against it rode the whole of the journey…'[21] It is well known that Crawshay was a betting man; in 1790 he had bet a German friend, Frederick von Reden of Berlin (who had previously ordered a steam-pumping engine from Homfray), about the completion date of the Glamorganshire Canal: '… the Canal goes beyond expectation… you will certainly lose your wager.' [22]

There is some ambiguity as to which of the Hill family acted as the stakeholder. Anthony Hill (1784-1862) is seen as the likely candidate because he later figures in seeking Brunel's interest in engineering a railway from Merthyr to Cardiff. However, Anthony Hill would only have been nineteen years old at the time of the wager and technically a minor[23]. It would be more appropriate for the father, Richard Hill, or his eldest son, Richard John Hill (1773-1845), to act as stakeholder and hold the bet made between Samuel Homfray and Richard Crawshay. Wilkins is also confused about the date of the wager run, giving the date as 12 February 1804. He even suggests it was 14 February 1804 in his *History of the Iron, Steel and Tinplate Trades*, although he is quoting the account from the journey that appeared in *The Cambrian* newspaper on 25 February 1804, which infers the date of the journey; 21 February 1804, as 'yesterday' in its account from Merthyr of 22 February.

For such a historic journey the amount of conflicting information is frustrating and the conflict goes beyond the actual parties to the wager and, indeed, if the wager was actually won. What is beyond dispute is that Trevithick was able to prove the worth of his engine by hauling five wagons carrying 10 tons of iron and seventy men from the Penydarren ironworks to the canal basin at Abercynon.[24] Unfortunately, although it proved the possibility of locomotive haulage, it was too heavy for the brittle cast-iron plates it ran on. There is evidence of at least one and possibly two more runs by the locomotive, although only in a southerly direction. Following that it was not used again as a locomotive engine but put into service as a stationary engine at Penydarren.

Richard Trevithick was to record the preparation for the journey and the historic episode itself in letters to Gilbert (still known as Giddy at this time):

'Penydarran February 15 1804
Mr Giddy

Sir,
Last saturday we lighted the fire in the Tram Waggon, and worked it without the wheels to try the engine, and monday we put it on the Tramroad. It worked very well and ran up hill and down with great ease, and was very managable. We have plenty of steam and power I expect to work it again tomorrow. Mr Homfray and the Gentleman I mentioned in my last, will be home tomorrow. The bet will not be determe'd until the middle of next Week at which time I shod be very happy to see you, I shall go down to Cornwall abt the end of next week or the beginning of the week after.

I am Sir Your very Hb Servt Richard Trevithick' [25]

'Penydarran Feby 22. 1804
Mr. Giddy

Sir,
Yesterday we proceeded on our journey with the engine we Carry'd ten tons of Iron five waggons, and 70 Men riding on them the whole of the journey, its above nine Miles, which we performd in 4 hours & 5 mints but we had to cutt down som trees and remove some Large rocks out of road the engine while working went near 5 miles pr. Hour, there was no water was put into the boiler from the time we started until we arrived at our journeys end, the Coals consumed was 2 Hund., on our return home abt 4 Miles from the shipping place of the Iron, one of the small bolts that fastened the axel to the boiler broak and left all the water out of the boiler which prevented the engine returning untill this evening, the Gentleman that bet five Hund.d Guineas

against it rid the whole of the journey with us, and is satisfyde that he have lost the bet, we shall continue to work on the road and shall take forty tons the next journey - the publick untill now calld mee a schemeing fellow but now their tone is much alterd, there is an engine orderd for the west India docks to travel itself from ship to ship to unload and to take up the goods to the upper floors of the store houses at the crane and in case of fire to force water on the store houses. The fire is to be kept constantly burning in the engine, so as to be ready at all times, Bolton & Watt have straind every nerve to get a bill in the house to stop these engines saying the lives of the publick are in danger by them and I have no dought but they wod have carryd their point if Mr. Homfray had not gone to London to prevent it, in consequence of which there is an engineer, from Woolwich warren orderd down and one from the Admiralty Office to inspect and make tryal of the strength of the matls and to prove that the steam gauges will admit the steam thro them in case the steam valve shod be fastened down. They are not to com down untill everything is compleat for those experements, you shale know of our future experements as fast as we get on with them.

I am Sir Your He St
Richd Trevithick'[26]

The Cambrian reported on the historic journey in its edition of 25 February 1804: '... It performed the journey without feeding the boiler or using any water and will travel with ease at the rate of 5 miles an hour'. The reporter was even moved to make the following prophesy:

> 'It is not doubted but that the number of horses in the kingdom will be very considerably reduced, and the machine in the hands of the present proprietors, will be made use of in a thousand instances never yet thought of for an engine'.

The same edition also referred to other examples of Trevithick's work: 'On Tuesday se'night one of Mr. Trevithick's steam engines was set to work on the colliery of Alexander Raby Esq of Llanelly' and '...another of Mr. Trevithick's engines, on the same principle, was exhibited in London last week...' In a letter to Gilbert from 5 January 1804, Trevithick complains about being at Swansea for the last four weeks: 'If I had not been at Swansea to put up the winding engine, the road-engine would have been at work long since...' Another high-pressure steam engine, although built by James Birch of the Abernant ironworks, was installed to drive an incline winding drum in the Vale of Neath early in 1805, but was found to be singularly unsuitable for its purpose.[27]

It is accepted that the locomotive did not fully complete the return journey under its own steam and, in his account, Rees Jones states that on a subsequent journey the engine made to Abercynon: '... she broke many of the tram plates; and before reaching the basin she ran off the road, and was brought back to Penydarren by horses'[28]. Trevithick refers to a breakdown on the return journey in his famous letter: '... on our return home, abt 4 miles from the Shipping place of the iron'. This would

George Yates' map of 1799 showing early Welsh tramroads ('waggonways') in the Neath district (see colour plate 8). (Courtesy of Glamorgan Record Office)

approximately be where the tramroad crossed the Merthyr Road. Indeed, the tramroad was to cross and recross the road several times. It would seem that in a further attempt to perform the journey and return in the same time that 'horses usually do', another fault befell the locomotive in the shape of an ineffective forcing pump.

It appears that only two more trips were run the full distance after the first. Another letter from Trevithick to Gilbert, writing from Stourbridge on 5 July 1804, states:

> '…the Tram engine have carryd two Loads of 10 Tons of Iron to the Shipping place since you left this place, but Mr. Hill says that he will not pay the bet because there was some of the tram plates that was in the tunnel removed so as to get the road in the middle of the arch, the first objection he started was that one shod go with the engine without any assistance which I performed myself without help, and now his objection is that the road is not in the same place as when the bet was made, I expect Mr. Homfray will be forced to take steps that will force him to pay…'[29]

Gilbert was present at Penydarren from the 12 March, and stayed for approximately ten days. On 10 July 1804, Homfray was to write to Gilbert:

Richard Trevithick (1771-1833) (From Francis Trevithick, Life of Richard Trevithick, *London, 1872.)*

A statue of Richard Trevithick, erected in front of Camborne Library in 1932, showing the side with the Penydarren locomotive panel. (SKJ photograph)

'... Trevithick went down the Tramroad twice since you left us, with 10 ton each time. & tho he took his load down, Mr Hill does not yet allow the 500 Gs. because he did not return again with the empty Trams in same time the Horses usually do. & this was owing to the little forcing pump not being quite right to feed the boyler & he was obliged to wait and fill with cold water - but this little defect is easily cured & no doubt but Mr Hill will be satisfied...'[30]

In his letter to Francis Trevithick in 1854, Thomas Ellis states that it made three journeys to the basin and, in coming up the third journey, it broke both axles.[31] This appears to have been the final setback and the reason why it was converted to a stationary engine. Although both Trevithick and Homfray talked about a final trial, this did not take place.

The Government or London engineers referred to by Trevithick had been invitedby Homfray to inspect the engine but, because of an accident that occurred to Homfray in early March 1804, this was not carried out. However, Francis Trevithick states in his biography that Mr Homfray's road coach was drawn on the tramroad by the engine with Homfray and a Government engineer in it. This engineer was: '... brought down for the purpose of examining and testing Trevithick's engine.'[32] This would have been shortly after the first run; however, a few pages later, Francis Trevithick talks about Homfray not being able to receive his '... friends from London, and the official tests were therefore deferred.'[33] The engineers likely to have been approached by Homfray were Simon Goodrich (machinist to the Navy Board) and/or Lt Cunningham of Woolwich, but there is no record of a visit from these engineers.[34]

Details of the locomotive itself are also sketchy because no authenticated drawings of the locomotive have survived; the illustration in Francis Trevithick's biography of his father is generally accepted as a representation of the engine only, and not an accurate record. A simple representation of the engine is that it consisted of a boiler containing a boiler flue which was basically a tube with a 'U'-bend. A single horizontal cylinder of an 8in-diameter was set into the end of the boiler. The piston of this cylinder drove a connecting rod, from a crosshead to a crankshaft, which transmitted power, via gears, to the wheels. On the other side of the crankshaft a large flywheel maintained regularity of motion. The wheels were flangeless so as to run on the cast-iron tram plates of the Merthyr Tramroad. Stories, such as the locomotive having a brick chimney (in Wilkins and other publications as late as 1930)[35] are also without foundation as the locomotive was constructed completely of iron, and Trevithick makes no mention of an incident that could form the basis of such a story, unless Wilkins is confused with the tunnel episode. A near-contemporary article (August 1805)[36] states that the chimney was of 'rolled iron' i.e. wrought-iron plates riveted together.

With regard to the origins of the materials used in the construction of the engine, there tends to be some contradiction. Thomas Ellis had been asked by Francis Trevithick in 1854 for evidence relating to Trevithick's achievements in South Wales. In a letter dated 22 June 1854, he states: '... from what I can understand the

Merthyr Tydfil Penydarren locomotive panel, one of the panels on the base of the Richard Trevithick statue in Camborne. (SKJ photograph)

engine was made in Cornwall, and put together in Merthyr…' In a further letter he mentions that: 'Rees Jones, who was aided in the fitting, and William Richards its driver are still alive'. The *Engineering* magazine of 27 March 1868 adds another member of Trevithick's assistants, John Roe, and confirms the part played by Rees Jones. Rees Jones himself made a statement which appeared in the *Mining Journal* on 2 October 1858: 'Most if not all the work of this locomotive was made at Penyderren. Richard Brown made the boiler and the smith work'. Richard Brown, formerly a master roller at Dowlais, built the boiler for Richard Trevithick's revolutionary steam locomotive in the Penydarren workshops over the winter of 1803/04.[37] John Steel was another mechanic that was invloved, a Newcastle man sent down to assist Trevithick with the Welsh engine. Steel then superintended the construction of the Newcastle engine. He became an early casualty of the high-pressure steam engine when a steam engine he was working on blew up, causing him to lose a leg.

That parts of the engine came from Cornwall cannot be verified, and it is more likely that the Coalbrookdale Co. supplied these parts: 'Harveys had no part in the manufacture of the Penyderren engine, all the parts were made at Coalbrookdale.'[38]

Trevithick begun the construction of a locomotive engine in the Coalbrookdale works in 1802 which appears to be very similar to the Penyderren engine but for some reason this engine, either never ran or was not completed. Trevithick went on from Penydarren to Newcastle where another engine similar to the Penydarren engine was constructed to run on a wooden edge railway. Francis Trevithick is of the opinion that parts of this engine were also made at Coalbrookdale.

For such a historic journey there is a lot of conflicting information; however, there is no doubt about the following: Richard Trevithick came to Merthyr to construct a high-pressure steam engine which had a wide variety of potential uses, including giving forced-draught to blast furnaces. Trevithick was at Penydarren from at least October 1803 and although he left to attend to other business, it appears to have been his base until around 23 July 1804. Writing to Gilbert from Coalbrookdale on 23 September 1804, Trevithick states: 'I left Wales about 8 weeks since'. On 1 October 1803, Trevithick wrote to Gilbert from Penydarren, stating: 'Mr Homfray of this place, has taken by me by the hand, and will carry both engines and the patent to the test'. From this letter it can be seen that Trevithick believed that Homfray would give him full support in trying out his engine and indeed, to take a share of the patent. The patent referred to by Trevithick was the one he had taken out jointly with Andrew Vivian for a steam engine propelling a carriage and so on on 24 March 1802. This was the first patent that Trevithick had taken out. Boulton & Watt had rigorously defended their extended patent, which did not expire until 1800, and James Watt treated any other form of steam engine as an attempt at patent evasion.

There was one order for a locomotive following the experiment, from Christopher Blackett of Wylam Colliery on Tyneside, but as the tramroad at the colliery had wooden rails, he did not take the engine. The engine was put to work as a fixed engine, 'to blow the iron down'.[39] In 1808 Blackett had the colliery lines re-laid with iron rails and wrote to Trevithick to supply a new engine for his new railway. It was too late; Trevithick had moved on and he told Blackett that he had: '... discontinued the business, was engaged with other pursuits, and could render no assistance'.[40] He felt he had taken the steam locomotive as far as he could, but had failed to achieve a breakthrough in terms of commercial success.

After Trevithick's departure there were no more locomotive journeys on the Merthyr Tramroad until 1832 (the year before Trevithick's death), when locomotives were reintroduced to the tramroad. After Penydarren, Trevithick's career was nothing if not eventful, and he was to disappear for eleven years to South America. In 1833 he was engaged at Hall's engineering works in Dartford, attempting to build a water-jet propulsion engine for ships. John Hall had commissioned Trevithick to build such a propulsion unit, for which he had taken out a patent in 1832. The engine kept exploding because of the high pressures involved, but local legend in Dartford has it that the jet engine pioneer, Frank Whittle, was influenced in his ideas for a jet engine by seeing Trevithick's patent of 1832, 'Application of Steam Power to Navigation and Locomotion', part of which (3) referred to 'Jet Propulsion of Vessels.'[41] After a long confinement to his bed he died from influenza, in a state of poverty, at the Royal Victoria and Bull Hotel, Dartford, on 22 April 1833. Having left no money for a burial, he faced the prospect of a pauper's funeral, but John Hall and others from the works contributed to the expenses of providing a decent burial, albeit in an unmarked grave in the churchyard of St Edmund, King and Martyr. The mechanics of Hall's works carried him to his grave. In a letter to his friend Gilbert, written less than a year before his death, he reflected on his life:

> '... I have been branded with folly and madness for attempting what the world calls impossibilities, and even from the great engineer, the late Mr. James Watt, who said ...that I deserved hanging for bringing into use the high-pressure engine. This so far has been my reward from the public; but should this be all, I shall be satisfied by the great secret pleasure and laudable pride that I feel in my own breast from having been the instrument of bringing forward and maturing new principles and new arrangements of boundless value to my country. However much I may be straitened in pecuniary circumstances, the great honour of being a useful subject can never be taken from me, which to me far exceeds riches.'[42]

Gilbert had, however, grown tired of this schemer – the inventor who had abandoned everything to chase his dreams, opting out of family and business responsibilities and switching from one project to another. In 1830 Gilbert was thoroughly exasperated by Trevithick and, while accepting that he was one of the most ingenious men in the world, thought: '... he has never done anything himself'.[43] He had realised Trevithick's major fault the year after Penydarren: 'No other fault is imputed to him than that, common to most Geniuses, of sacrificing in some instances routine employment and certain gain to prospects of Fame or great Wealth.'

It is possible to walk most of the route of the tramroad today, with much of the route forming part of the Taff Trail. The start of Trevithick's journey on the tramroad is marked by a stone monument bearing the following inscription: 'This monument built of chair stones and rails recovered from the old Merthyr Tramroad was erected by the people of Merthyr Tydfil with the help of the Trevithick Centenary Commemoration Fund, November 1933.'

Today the monument is set off with a model of the locomotive, which was part of the 1933 design but has only recently been completed. From the monument the course of the tramroad crosses Nant Morlais and runs along lanes known as Tramroad Side North and Tramroad Side South, passing the side of the former Miners Hall (a building with Brunel connections) and continuing past the 'Glove & Shears' public house. Approaching the area that was once occupied by the Plymouth ironworks can be found a major surviving structure on the tramroad in the shape of a tunnel, whose southern entrance has been restored, and is today known as 'Trevithick's Tunnel'.

Continuing southwards from the tunnel to Troedyrhiw, much of the route of the tramroad has been obliterated by land reclamation, the construction of the A4060 road and commercial and industrial development. It is still possible to follow the approximate route, passing behind the present-day Linde factory. The tramroad enters the village of Troedyrhiw behind a row of old houses and the route can be followed right through the village. Evidence, in the form of the stone sleepers, can still be seen in places. At the southern end of the village the tramroad crosses the A4054 (the old Cardiff Road) and passes a small caravan site before recrossing the

TREVITHICK'S LONDON RAILWAY AND LOCOMOTIVE OF 1808. [W. J. Welch.]

Catch-Me-Who-Can, *July and August 1808. Trevithick ran his locomotive* Catch-Me-Who-Can *on a circular track near Euston Square in London. It appealed mainly to thrill seekers, who paid a shilling a head to be transported at speeds of up to 12mph. It was Trevithick's last attempt to promote the worth of his steam locomotive. (From Francis Trevithick,* Life of Richard Trevithick, *London, 1872)*

A section of Merthyr Tramroad and two wagons. This photograph was taken to illustrate the length of plateway and wagons loaned to the World's Columbian Exhibition at Chicago in 1893. The exhibition included a transportation section covering the development of the steam locomotive. The individual is possibly Thomas Henry Bailey (1857-1939), who was brought in to manage the colliery interests of the Plymouth iron works, which had been put up for sale in 1882. The enterprise became known as Hill's Plymouth Co. Ltd. (Courtesy of the Institution of Civil Engineers)

An upstream view of the tramroad and widening work in progress at Goitre Coed Viaduct, Quakers Yard. Taken in 1862, it is the earliest photograph showing the Merthyr Tramroad in operation with a southbound horse-drawn train just having given way to a northbound train. Goitre Coed Viaduct is in the process of being widened. (Joseph Collings photograph/John Minnis collection)

The Penydarren monument at Pontmorlais, Merthyr. Originally built for the 1933 anniversary marking the centenary of Trevithick's death, it was rededicated on 21 February 2004. The replica of the locomotive was only added in recent years. (SKJ photograph, taken on 21 February 2004)

road. This section follows the original line almost exactly, although the tramway originally crossed the parish road at Pontygwaith at ground level and was realigned when the overbridge was built. A bridge crossing the tramroad here carries a very steep road down the side of the valley to the site of an ancient iron furnace. The name Pontygwaith comes from this association and means 'works bridge'. It refers to the original bridge crossing the river Taff. Examples of the original stone sleeper blocks can be more readily found, in situ and with rail imprints, south of this bridge.

Below this point the tramroad passes under three railway viaducts. With only the piers remaining of the first two, approaching Quakers Yard, the tramroad passes under the Goitre Coed Viaduct of the Taff Vale Railway. Beyond it, the tramroad crosses the Taff by means of a masonry bridge of 63ft-span, which was built to replace an earlier structure. The original wooden bridge collapsed while carrying a train of trams in 1815, and was rebuilt as an elliptical stone arch.[44] Just over ½ mile downstream, a second bridge recrosses the Taff, also rebuilt in stone at the same time, and carries the tramroad over the Taff. Samuel Downing, the resident engineer for the Goitre Coed Viaduct on the Taff Vale Railway, made the following comment about the need for these bridges, in 1851: 'Even the Merthyr Tramroad, which acted as an auxiliary to the canal, on the upper part of its course, where it had not the command of the whole water of the river, had, as may be seen, two considerable bridges over the Taff to avoid these disadvantages'[45]

The journey ends at Navigation, now known as Abercynon, where a memorial has been erected outside Abercynon fire station, close to the Navigation Hotel, marking the end of Trevithick's historic locomotive run at the 'shipping place of the Iron.' Over the course of the journey existing memorials have been rededicated and new works commissioned in 2004, as part of the bicentenary events celebrating that epoch-making journey.[46]

Chapter 3 Notes

1. Royal Institution of Cornwall (RIC), Truro. Enys collection Trevithick papers. Trevithick to Giddy (Gilbert) 22 February 1804. Quoted (with corrected spelling) in: Trevithick, Richard. (1872). pp.161-62, (Vol.1). *Life of Richard Trevithick with an Account of his Inventions*, Vols 1 & 2, London: E. & F.N. Spon.
2. Trevithick & Vivian Patent No.2599, 24 March 1802.
3. Dickinson, H.W, & Titley, T.A. (1934), p.43. *Richard Trevithick*, Cambridge University Press.
4. Trevithick, Richard. (1872). pp.125-27. (Vol.1).
5. Trevithick, Richard. (1872). p.162. (Vol.1).
6. Galloway, Elijah. (1831). pp.42-43. *History and Progress of the Steam Engine*. London: Thomas Kelly.
7. Galloway, Elijah. (1831). p.43.
8. Trevithick & Vivian Patent No. 2599, 24 March 1802.
9. RIC, Gilbert to J.S. Enys, 29 April 1839, quoted in Todd, A.C. (1967). p.83. *Beyond the Blaze: A Biography of Davies Gilbert*. Truro: D. Bradford Barton.
10. *The Cambrian*, 25 February 1804.
11. Lewis, M.J.T. (1970). p.331. *Early Wooded Railways*. London: Routledge & Kegan Paul.
12. RIC, Trevithick to Giddy (Gilbert), 1 October 1803.
13. RIC, Trevithick to Giddy (Gilbert), 20 February 1804, quoted in Todd, A.C. (1967). p.85.
14. Todd, A.C. (1967). p.86. Diary 22 March 1804.
15. RIC, Wood to Giddy (Gilbert), 21 March 1804. Todd, A.C. (1967). pp.86 and 92.
16. Trevithick and the Merthyr Tramroad', Stanley Mercer.
17. *The South Coal Trade* and *The History of Merthyr*
18. RIC, Letter by Trevithick to Giddy (Gilbert) 22 February 1804. Quoted (with corrected spelling) in: *Trevithick, Richard*. (1872). pp.161-62.
19. *Mining Journal* for 2 October 1858
20. Rolt, L .T.C. (1960). *The Cornish Giant*. Lutterworth Press.
21. RIC, Trevithick to Giddy (Gilbert) 22 February 1804. Quoted (with corrected spelling) in: *Trevithick, Richard*. (1872). pp. 161-2.
22. Rowson, Stephen & Wright, Ian L. (2001). p.31. *The Glamorganshire and Aberdare Canals*. Vol.1. Lydney: Black Dwarf Publications. See also Henderson, W.O. (1958). pp.1-20, *The State and the Industrial Revolution in Prussia 1740-1870*, Liverpool.
23. Thomas, D.R. (nd 1948?), *The Penydarren Locomotive*.
24. Stanley Mercer estimated the weight of Trevithick's locomotive as follows; Engine 5 tons, Iron 11 and a quarter tons, wagons 4 tons, men say 4 and three quarter tons, making a total of 25 tons (Trevithick's ton was 'long weight' which was considerably heavier than the modern ton).
25. RIC, Trevithick to Giddy, 15 February 1804.
26. RIC, Trevithick to Giddy, 22 February 1804.
27. Mear, John F. (1999). p.40. *Aberdare, the Railways and the Tramroads*. Aberdare: pub., by the author.
28. *Mining Journal* for 2 October 1858.
29. RIC, Trevithick to Giddy, 5 July 1804.
30. RIC, Homfray to Giddy, 10 July 1804.
31. Trevithick, Richard. (1872). p.174, (Vol.1). Thomas Ellis to Francis Trevithick, 15 May 1854.
32. Trevithick, Richard. (1872). p.163 (Vol.1).
33. Trevithick, Richard. (1872). p.168 (Vol.1).
34. Records at the Public Record Office (W.O.55) relating to the Ordnance Office and Woolwich at the relevant period appear to contain no references to Merthyr or the tramroad. Many of the entries, reports etc. relate to inspection of defence works such as Milford Haven. Dickinson, H.W & Titley, T.A. (1934) put forward the engineers Simon Goodrich and/or Lt Cunningham.

35 Teachers Association. (1930), *Story of Merthyr Tydfil*, Merthyr.
36 *Nicholson's Journal of Philosophy*, August 1805.
37 Evans, Chris. (1993), p.76. *The Labyrinth of Flames, Work and Social Conflict in Early Industrial Merthyr Tydfil*. Cardiff: University of Wales Press.
38 Vale, Edmund. (1966), *The Harveys of Hayle*.
39 Trevithick to Gilbert, letter dated 5 July 1804, quoted in Burton, Anthony. (2000). p.95. *Richard Trevithick Giant of Steam*. London: Aurum Press.
40 Burton, Anthony. (2000) p.95.
41 *History of Holy Trinity Church*, Dartford.N.D.
42 Trevithick, Richard. (1872). pp. 395-6, (Vol 1). Trevithick to Gilbert 14 May 1832.
43 Todd, A C. (1967). p.111. (1805).
44 Owen Jones, Stuart. (1981). p.14. *The Penydarren Locomotive*. Cardiff: National Museum of Wales.
45 Downing, Samuel. (1851). p.25-26. *Description of the Curved Viaduct at Goed-re-Coed near Quakers Yard, Taff Vale Railway*. Trans. Institution of Civil Engineers of Ireland. Vol.4, Part 1.
46 *Archive*, Issue 44, December 2004. Jones, Stephen K. (2004), pp.57-64. 1804: *The Year of Trevithick's Dragon*. Includes photographs of the memorials.

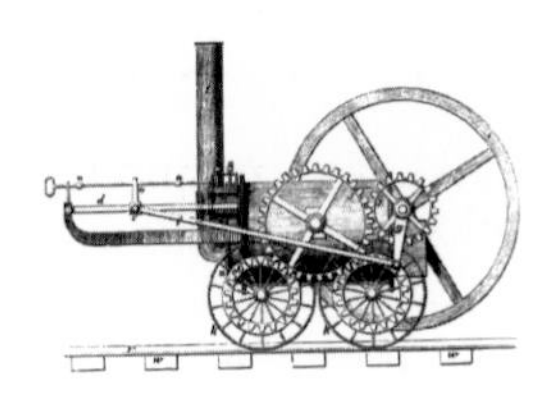

4
Grooves Of Change
'Tramways Ran in Every Direction...'[1]

In the years after Trevithick's run, the development of the steam locomotive was to pass to others such as John Blenkinsop and Matthew Murray. Like Trevithick, Blenkinsop and Murray also introduced a steam engine to an existing railway. In this case it was the Middleton Colliery Railway, which was built as a wooden waggonway in 1758 and ran for just over 3 miles from Middleton, to supply coal to the town and neighbourhood of Leeds. In 1808 the wooden rails began to be replaced by iron and, in 1812, the manager of the colliery, John Blenkinsop, employed the engineer Matthew Murray of Tyneside, who was then a partner in the Leeds foundry of Fenton, Murray & Wood, to build a steam locomotive to work on the colliery. They had learnt from the problems encountered by Trevithick, namely that the weight of the locomotive was too heavy for the rails it ran on. To overcome this a third rail was laid on the track; this was a toothed rail which engaged with a cog wheel on the locomotive and allowed a lighter engine to be used without losing traction. It was the first rack-and-pinion railway, and the first locomotive railway to be commercially successful. The colliery proprietors paid a royalty of £30 for each of the two locomotives built, in respect of Trevithick's patent, but Trevithick was not to benefit as by then he had sold his share in the patent. While other colliery locomotives were to be built by Murray and Blenkinsop over the next two years, and one possibly at Nantyglo[2], the Middleton rack-and-pinion system was not followed with any of these. Notable among these locomotives was the Killingworth engine built by George Stephenson, for which a joint patent was taken out with Ralph Dodd.

Ironically, the next authenticated reference to a steam locomotive in Wales relates to one supplied by George Stephenson in 1819. This locomotive ran on the railway that served Scott's Pit near White Rock, Llansamlet.[3] There was a strong effort by the owners to introduce the mining and other practices of the north-east of England to the area, including a Killingworth manager and the laying of an edge railway in an area dominated by tramroads. Stephenson was to go on to build the world's first passenger railway; the Stockton and Darlington Railway (SDR) which opened in 1825. It had taken twenty-one years following Trevithick's historic run to establish the steam locomotive as a credible form of motive power. There was a more direct link with the SDR, and one that went back to the Merthyr Tramroad, through the engineer George Overton. Overton, as one of the foremost tramroad builders of his day, was a good choice to advise on the building of a railway. He was also related to Jeremiah Cairns, the agent of Thomas Meynell, one of the principal figures behind the Stockton and Darlington proposals.[4] Jeremiah Cairns played a significant role in the development of the SDR and in 1838, Cairns, now living in Newport, was to contribute to the TVR story by assisting in the visit by George Bush, Thomas Powell, Walter Coffin and Anthony Hill, '...with the purpose of looking over the SDR and obtaining much information as possible...'[5] (see Chapter 7).

There had been a number of proposals before Overton was consulted. Indeed, the original proposal dated back to 1768 for a canal to take the coal mined in the pits around west Durham to be shipped at Stockton, the highest navigable point on the Tees. The celebrated canal engineer, James Brindley, carried out a survey and estimated the cost to be £63,722. Nothing further transpired until 1810, when it was raised again at a dinner at Stockton to celebrate the opening of a short navigational cut, some 210 yards long, on the Tees. At the dinner the question of a canal came up again, and the suggestion of a tramroad was also put forward. Edward Pease, the Quaker merchant, was present, and a committee was one of the outcomes. The opinion of John Rennie was sought and his report, completed in 1813, was in favour of a canal, estimated to cost £205,618. This was the highest estimate so far and the proposal lay dormant until 1818. Another canal survey in that year by George Leather was commissioned by Stockton merchants, including Christopher Tennant, and favoured a more direct route which would avoid Darlington. Meynell believed that his interests, as a landowner in Yarm, would also be bypassed. Meynell dismissed the Stockton canal as 'a wild scheme' and approached Overton with a view to commissioning a survey which would give justification for a railroad, making it clear that he was in favour of such a proposal.

To assist him, Overton took the surveyor David Davies with him, and a suitable course for a railroad, running from Stockton to Etherley, was found. Overton's estimate of £124,000 covered a main line of 35 miles and eight branches totalling 16 miles.[6] It was the first survey for the SDR, although Overton also surveyed a route for a canal which would pass through Darlington. He presented his report to the Darlington committee (chaired by Thomas Meynell) on 29 September 1818. In order to get the scheme into the next parliamentary session, the plans were hurriedly prepared to meet the deadline for the 1818-19 session of Parliament, and deposited by the end of the month, even though no decision had been taken about whether there should be a canal or a tramroad. The decision to build a railway was taken two weeks later.

There was, however, some uncertainty about the soundness of the proposal together with growing opposition from landowners, such as the Earl of Darlington (later the Duke of Cleveland), over the line going through their land. To consider alternatives, other engineers were consulted, referring back to John Rennie and bringing in Robert Stevenson, the lighthouse engineer. Rennie and Stevenson had crossed swords over the Bell Rock Lighthouse and the former was still in favour of a canal. Public conflict was avoided when Rennie replied to the SDR committee's request on 26 December 1818, stating that he was far too busy with other public works '... of infinitely greater magnitude and importance than the Darlington Railway'.[7] Stevenson's initial advice, apart from not calling in any other engineer (presumably a reference to Rennie) was to press on as quickly as possible with the present survey. He was to later suggest another route, but this was not taken forward. In May 1821 Robert Stevenson was to write to Overton in reply to his request for information on his engineering experience[8]. The letter related to work Stevenson was reporting on to John Guest about improving communications into South Wales, by roads and ferries, rather than his involvement in the Stockton & Darlington.[9]

The Bill was defeated, but by a narrow majority that encouraged the Quaker merchants led by Pease to attempt a second attempt. Meynell was one of the few influential supporters who was not a Quaker, and virtually the only landowner who approved of the railroad proposal and, in 1819, he subscribed £3,000 towards the project. On 12 February 1820, he chaired a meeting at the 'George and Dragon Inn' in Yarm, which was to ultimately lead to the SDR receiving parliamentary approval.

Overton was reluctant to undertake further work, particularly with the committee seeking the advice of other engineers and having to put up with political squabbling by the various factions involved, but he eventually agreed to prepare a second survey for a guaranteed fee of £120.[10] This survey avoided Lord Darlington's fox-hunting reserves and, despite the consequences of pushing the proposed line across the landed interests of other peers, it managed to enlist wider support for a second Bill. The illness and death of George III led to over a year's delay, but the Bill received its Royal Assent on 19 April 1821. It was also on that date that George Stephenson, with Nicholas Wood, met with Edward Pease at his home in Darlington. Despite his personal connections, and that he had himself subscribed £2,000 to the SDR, Overton found that he did not have the full support of Pease.

Overton's opinion was that the line should be built as a tramroad and operated by horses because he believed that locomotive haulage was fraught with problems, a view that probably stemmed from observing the event of 1804: '... I have seen Trevithick's Engine heading trains 20 years ago...'[11] Overton made his opinions known that it should be a tramroad, which was relayed in a letter from Jeremiah Cairns of the SDR to Guest on the '... Subject of Railroads and Tramroads, that I determined upon appealing to your experience for your opinion'[12]. Cairns' preference was for a tramroad although he acknowledges that others, such as William Jessop, (in an annexed copy of a letter also sent) believe; '... that the edge Rail Ways are very superior to the Tramroads of this County [Derbyshire] and Wales'[13] and mentions that 'Mr Frere has also written and given the preference to the Rail Way but he says that there is not much difference' (This would appear to be Edward Frere, a partner in the Clydach ironworks in Breconshire, whose second son George Edward was to become one of Brunel's chief assistants). Cairns relates to Guest in his letter that '... Mr Overton has given his sentiments in decidedly in preference to the Tramroad'. Cairns concurs with this through his own experience in Wales: '... for I recall the Limestone Railroad to Aberdare being replaced by a Tramroad'.

Edward Pease had also sought advice from engineers, such as Joseph Tregelles Price of the Neath Abbey ironworks, on the merits of tramroad versus railroad some three years earlier. In his reply, dated 22 December 1818, Price demonstrates a preference for tram plates, but mentions: '... there is at this time a line of Railroad making within 4 miles of this place by a person who, as Surveyor etc., has had a great deal to do with tramroads. He is, however, doing it by direction of his Principal'.14 This is part of the evidence for Scott's Pit Line being an edge railway and which contradicts William Jessop's opinion from 1821 that: 'I do not know a single edge Rail Way in Wales & I do not believe they are known here, except the old Wooden Waggon Roads which were in use before Iron was applied...'[15]

Overton also dismissed the idea of railways forming part of a national system through observations made in a book he published in 1825[16]. Such views were contrary to those of Stephenson, whom Pease was eventually to consult. Stephenson's opinion would resolve the tramroad versus edge railway argument, and influence Pease's views on locomotive haulage. Pease sought Stephenson's examination of Overton's survey with a view to ascertaining whether the construction of such a line was practical, and if the route could be improved, in terms of reducing landowner opposition and providing easier gradients for locomotives. Meynell was, naturally, against Stephenson resurveying the route as the route they had was good enough to have been passed by Parliament, but his opposition was eventually won round.

On 12 May 1821, Meynell was appointed the first chairman of the new railway company, and the following day he wrote a letter to the Dowlais Iron Co. which is not addressed to Guest or any other individual, despite his connections with Overton. The letter says that he: '... will esteem it a great favour if you will do me the honour to communicate to me your opinion of the comparative value and advantages of tramroads and railroads as constructed on the most improved principles, also the grounds of such preference and particularly the quantum of work one horse is capable of...'[17]

In July 1821 Stephenson was appointed to carry out a new survey, and it was completed by January 1822. He had managed to shorten Overton's line by just under 4 miles, and to improve the gradients of much of the route. The committee's satisfaction with this new route resulted in Stephenson's appointment as engineer to the SDR. Several months later, in July 1822, Overton wrote to John Guest: '... I do not know whether I told you that Stephenson is the engineer who had succeeded me as Engineer or rather superceded me as Engineer to the Darlington Road.'[18] Overton goes on to say:

> '... After Stephenson's appointment I was disposed, although unknown to him, to be Friendly, & on some points we differ'd. I invited him to call upon me to discuss them, should he come to Wales, but when his report was publish'd & sent to me, I wrote a Letter back, tacitly accusing him of stating falsehoods, in fact I offered to bet him £100 that he could not support his assertion in speaking of my Line, & some alterations he recommended. I do not know that my Letter has been shown to him, but it is very probable that it was ...'

George Stephenson, 1781-1848. (SKJ collection)

In this letter he also asks if John Guest will support his testimony to the jury of the SDR to adopt his preferred choice of tram plate and sleeper. Leaving aside the issue of the rail, what to do about supporting the rail was a problem that Stephenson had to address. Stephenson could not use transverse sleepers, as he needed to lay his rails on blocks either side to allow an open path for horse-drawn traffic, which would still be used on the line. However, by the end of 1821, Stephenson had already decided to use malleable iron rails of Birkinshaw's patent. Jeremiah Cairns wrote, on behalf of the SDR, to Guest on 1 November 1821 with a specification for Dowlais to supply wrought-iron rails (within the description of Birkinshaw's patent)[19]. John Birkinshaw had patented his wrought-iron edge rail in 1820 and, following its successful adoption on the SDR, it would also be used on the Liverpool & Manchester Railway. Despite this type of endorsement, the arguments over the merits of wrought-iron edge rails compared with cast-iron edge rails or plate rails continued until 1830, when the superiority of wrought iron for higher speeds finally won the day.[20]

The power to use locomotives, and to incorporate the diversions resurveyed by Stephenson, were sought in a new Bill submitted to the 1822-23 session even though work had started (under the powers of the 1821 Act) with a formal rail-laying ceremony on 13 May 1822. When the line opened in 1825, the first train was worked by *Locomotion*, a locomotive manufactured by Stephenson's purpose-built locomotive works, Robert Stephenson & Co.

Around this time Brunel was acting in an engineering capacity on behalf of his father. In 1825 Marc Isambard Brunel had become involved with the proposed Grosvenor Bridge at Chester. Arriving in Chester on 9 February 1826, he presented the report and estimates prepared by his father to the Grosvenor Bridge commissioners. Marc Isambard Brunel was to reinforce his confidence in his son, representing him in a letter: '... I am not afraid of his committing himself by any injudicious observations, or in any calculations that might be necessary to prove the practicability of my plan of employing brick for an arch of the proposed scale.'[21]

Brunel took the opportunity to travel from Chester to Bangor by the mail coach, returning to Chester on 11 February 1826.[22] It is more than likely that he did this to see, at first hand, Telford's suspension bridge across the Menai Straits. Opened to great fanfare some ten days earlier on 30 January 1826, Telford's bridge was then the longest suspension span in the world. Along with the road improvements engineered by Telford, the Menai Bridge reduced the travelling time between London to Holyhead from 36 hours to 27 hours. Marc Isambard Brunel was to withdraw as a candidate to build the Grosvenor Bridge in July 1826, due to difficulties with the Thames Tunnel, the bridge being built to the designs of Thomas Harrison. When the bridge was opened in 1834 it was, with a single span of 200ft, the world's longest masonry span, and is still the longest in Britain.

A suspension bridge was to figure in the Middlesbrough extension of the SDR, which opened in 1830. It was designed by Capt. Sir Samuel Brown RN who, as well as pioneering the development of the chain suspension bridge in Britain, had, virtually single-handed, been responsible for the introduction of iron chain cable for ships. In January of that year, Brunel sought the position of engineer on the Newcastle & Carlisle Railway, but he was in Stephenson territory, and was unsuccessful. Other setbacks in securing engineering positions followed, and in 1833 he was working on the Fossdyke Navigation.

Back at Merthyr there was still an extensive network of tramroads; Gilbert had been impressed enough to seek advice on building a 'railroad' from Homfray's manager, and this eventually led to Lord de Dunstanville laying the: '... first iron of the Railroad...' for the Poldice Tramroad in October 1809.[23] Michael Faraday (1791-1867) was to be similarly impressed on his visit to Wales in 1819, spending several days in Merthyr as the guest of John Guest in July that year: '... Mr Guest would not allow me to go back to the Inn, I slept very soundly, though the fires from the hills kept up a continual roar; which though softened by the distance, had still a very impressive effect.'[24] Even Dowlais House was a noisy and sometimes dangerous place. Shortly after her marriage to John Guest, Lady Charlotte had experienced an even more 'impressive effect' when the boiler of the new forge exploded, bringing

RATES OF TONNAGE,

Commencing on the 1st. of August, 1830.

No.		D.	
No. 1.	For all Coals and Cinders, for home consumption, carried on the Main Line or any of the Branches	2¼	Per Ton per Mile.
2.	For all Coals put on board any Vessel in the Port of STOCKTON-UPON-TEES, to be delivered at HARTLEPOOL, or SALTBURN	1½	Per Ton per Mile.
3.	For all Lime, or Stone, put on Board any Vessel, in the Port of STOCKTON-UPON-TEES, to be delivered at HARTLEPOOL, or SALTBURN	1	Per Ton per Mile.
4.	For all Stones and Gravel, to be used for the making or repairing of Public or Private Roads, which may be carried on the Main Line or Branches	0½	Per Ton per Mile.
5.	For all Marl, Sand, Clay, and Manure carried on Ditto	1	Per Ton per Mile.
6.	For all Lime carried on Ditto	1¾	Per Ton per Mile.
7.	For all rough Rubble-Stone for Building, Ashlar and Hewn Stone, Bricks and Tiles carried on Ditto	1½	Per Ton per Mile.
8.	For all Slate, Lead in Pigs, or Sheets, Bark and Timber; also all Meal, Flour, Grain, Straw and Hay, Bar and Pig Iron, Staves and Deals carried on the Main Line and Branches	1½	Per Ton per Mile.
9.	For all Goods, Commodities and Merchandize not before specified, carried on Ditto	2½	Per Ton per Mile.
10.	For all Coals put on Board any Ship or Vessel, Lime, Stone, Goods, Wares and Merchandize, or other Articles deposited in or upon, or loaded or unloaded at any Depot, Shoot, Staith, Wharf, Warehouse or Landing Place belonging to the Company	1	Per Ton.
11.	For all Coals or Cinders, which shall be Shipped on Board any Vessel at or near STOCKTON, for the purpose of exportation, or for the purposes of navigation	0½	Per Ton per Mile.
12.	For all Lime and Stone, which shall be Shipped on Board any Vessel in the Port of STOCKTON-UPON-TEES, for the purpose of exportation	1	Per Ton per Mile.
13.	For all Coals and Cinders, which shall be put on Board any Vessel at or near MIDDLESBROUGH, for the purpose of exportation	0½	Per Ton per Mile.
14.	For Ditto Ditto, so far as the same shall pass on the Middlesbrough Branch Railway, the additional sum of	1	Per Ton per Mile.
15.	For all Coals, Lime, Stone and Cinders, which shall be carried or conveyed upon, or over the Bridge of Suspension erected across the River Tees	2	Per Ton.
16.	For all Articles, Matters, and Things, for which a Tonnage is hereinbefore directed to be paid, which shall pass the Inclined Planes, worked by the Permanent Engines	6	Per Ton.
17.	For all the Articles, Matters, and Things, for which a Tonnage is hereinbefore directed to be paid, descending any one Inclined Plane worked by the Permanent Engines, and which shall not have ascended any such Plane	2	Per Ton.
18.	For every Chariot, Coach, or other Carriage, which shall be used for the conveyance of Passengers on the Main Line, or any of the Branches	3	Per Mile.
19.	For Ditto, Ditto, on Sundays	6	Per Mile.

PERCIVAL TULLY, COLLECTOR at Darlington.
THOMAS LANGSTAFF, DITTO at Croft.
DITTO, at Yarm.
DITTO, at Stockton.

13th July. **Signed J. PEASE, Jun., Chairman.**

Stockton & Darlington rates of tonnage. (Courtesy of the Institution of Civil Engineers) down a 120ft-high chimney stack which killed three workmen. Unlike Faraday, Lady Charlotte found it difficult to: '… convey any idea of the crash and terror with which it was accomplished.' Iron fragments speared into the wall of the house, and bricks and stones rained down: '… a large brick was found in my bed which I had not long left.'[25]

Before arriving at Dowlais, Faraday had rambled around Cardiff Castle and gardens and walked along the canal to the sea lock, commenting that;

> 'All the iron from Merthyr, Dowlais and that neighbourhood is brought to this port, there being both a canal and a tramway from those places to the harbour. There are flood gates at the termination of the canal which keep it constantly full of water… much haematite (I believe from Lancashire) was lying on the wharf.'[26]

Presumably, Faraday was aware that the 'tramway' did not run the whole length but was a link in the transport system. Travelling up to Merthyr the following morning, Faraday noticed the change in the countryside: 'Tramways ran in every direction, and every now and then a range of thirty or forty trams laden with coal, or ore, or limestone, illustrated the advantage of this mode of conveyance.' He makes a detailed examination of the tramroads serving the works:

> 'There are more than 50 miles of tram-roads in the Dowlais works only. The trams, from which the road takes its name, are small wagons, sometimes of wood and sometimes of iron, which run on four low cast-iron wheels. There are no shafts or traces to them, but a strong bar runs down the middle, at each end of which a large hole is made. Iron loops about two feet long are made use of to connect from hole to hole and thus 20 or 30 trams are joined together in a few moments without too much trouble. Horses are attached to them by a pair of shifting shafts which hook into the holes before-mentioned; or sometimes they are made fast simply by chains. The tram-roads are formed of plates of iron about 2 feet or 30 inches long, 3 inches wide and of sufficient thickness to bear the weight and jolting of the trams. A ledge is raised on one side of them as a guide to the wheels. When laid down, spars of wood or stones are placed in the earth at such distances that the end of the tram irons may rest upon them and then they are arranged end to end in two lines, distant from each other so that they may fit the wheels of the trams; they are retained in their places by spikes passing through holes in the end of each piece …'[27]

Faraday also describes how one of the tramroads up and down a hill was operated by cylinder instead of horses (self-acting incline). By this method, the descending tram could take down more than 6 tons in weight, the ascending tram obviously being considerably lighter in order to balance the load. On an ordinary stretch of line, such as the Merthyr Tramroad, it was reckoned that one horse was able to pull a train of up to 10 tons downhill, and bring the empty wagons back, which was a round trip of 19 miles. Tolls were calculated on a ton-per-mile basis, although there were problems with unauthorised traffic on the Merthyr Tramroad, such as people riding horses and people walking the line, for which the company brought in fines. In 1812, the Merthyr Trust Renewal Act included a clause (xxxiv): '... A fine of £2 will be imposed on persons who shall ride, go, pass or travel with a horse, mare, gelding or other beast along the said tramroad except for the purposes of drawing a tram and wagons.' [28]

With all the events that had gone on since 1804, a fresh attempt at using locomotives to draw wagons in the Merthyr area was to be made at the end of the 1820s. William Forman of the Penydarren works was to order a locomotive from Robert Stephenson & Co. in December 1828.[29] The engine was originally intended to be an 0-6-0 engine with twin vertical boilers, similar to a locomotive then being built for the Liverpool & Manchester Railway, but fitted with plateway wheels instead of flanged wheels, to run on a 3ft-gauge. Changes were made to the design and the engine, and Forman's engine (later known as the *Eclipse*) now had a horizontal boiler and was to cost £375, with additional charges for the packing case and shipment. Two engines were despatched from the Stephenson's Newcastle works on 18 July 1829, and sent by coastal vessel around to Newport where they were unloaded at the Tredegar Iron Co.'s wharf. One locomotive, *Britannia*, was destined for the Tredegar ironworks while the other was taken on to Penydarren by road.

William Hedley's Puffing Billy *as an exhibit at the Patent Museum (now the Science Museum, London). Note the fish belly pattern rails. (SKJ collection)*

The news of a steam locomotive returning to Merthyr would have spread rapidly and no doubt the Crawshays would have been aware of such developments. William Crawshay II (1788-1867) was living at Cyfarthfa and running the works, although not without interference from his father, and it appears that he was so keen to be involved in this resurgence of the steam railway that he persuaded Goldsworthy Gurney (1793-1875) to bring one of his steam carriages to Wales. But, before this episode, William II had attempted to develop his own steam locomotives as there is a reference to William I and his engineer, William Williams, having built an engine that they considered entering in the Rainhill Trials in 1829. Sadly, little is known of this or of a locomotive offered to the Liverpool & Manchester Railway in 1830.[30] It is possible that William II was considering manufacturing steam locomotives and knew locomotive engineers such as John Braithwaite, who had an arrangement with Gurney to manufacture some of his engines.[31] William Crawshay I appears to have put an end to this by writing to his son telling him that:

> '... I thot we were equal to all our wants in that way at home and that Will Wms was to have been a racer with one of his own made engines at the Liverpool Race. Sad loss must have been accrued by that engine and other such experiments.'[32]

A reply from his son was followed by a further letter by William I, reinforcing his desire for not going outside for work, or engaging in frivolous activities such as ordering Gurney's engines, '... To make engines in Wales to run races in Lancashire is not my wish or interest or profession. Nor do I think otherwise of having Gurney and his engine in Wales.'[33] The reference to races in Lancashire referred to the Rainhill Trials of the Liverpool & Manchester Railway which took place in October 1829. John Braithwaite, with John Ericsson, had entered their locomotive *Novelty* in the trials. Gurney was one of the pioneers of steam road travel and had designed the first of his steam 'drags' or tractor units for pulling carriages in 1828. William II wanted to use Gurney's steam drag, which was pulled by horses from London to Cyfarthfa, on his new Hirwaun Common tramroad, but was to try it out first on a short stretch of plateway laid down at the Cyfarthfa works. The wheels were changed to a cast-iron set and a carriage with a test weight of iron (23 tons) was attached. It proved to be too much for some of the temporary tram plates. Gurney then drove the drag, still fitted with tramroad wheels, up the incline of the common road and the drive to Cyfarthfa Castle, where he turned in a figure of eight in front of William II's castellated mansion.[34] The following day it was taken to Hirwaun, and placed on the tramroad between the Hirwaun ironworks and Park Pit. It drew six carriages and a load in excess of 20 tons and upwards of 100 persons (possibly picking up a further 47 during the course of the journey) along a 3-mile stretch in 39 minutes.[35] It returned with the same load in 32 minutes. William II talks about subsequent days being devoted to various private experiments with the result that:

> '... the Engine drew a load of 21½ tons upon the same 3 miles in 19 minutes and returned in 18 minutes... I shall only call the attention of your readers to the

following points: - In all cases named, Mr Gurney's Engine has drawn from 15 to 16½ times her own weight, upon a level road and has produced more than sufficient steam for the purpose...'[36]

William II was more favourably disposed to the steam locomotive than his father, ordering a second steam drag in November 1830. One of the engines is believed to have worked at Hirwaun for nine years.

Meanwhile, Forman's engine was put to work on the 3ft-gauge Penydarren works network, a duty it performed for two years. If this was a testing process then it was successful because the engine was returned in 1832 to Stephensons for conversion to the 4ft 4in-gauge to enable it to run on the Merthyr Tramroad. It was not just the gauge that was changed; a considerable amount of rebuilding was also carried out, such as boiler tubing and a new frame, and *Eclipse* emerged more like a standard four-coupled Stephenson 'Planet' type locomotive, only retaining inclined cylinders. What was almost a new locomotive, officially called *Eclipse*, began work on the Merthyr Tramroad on 22 June 1832. *The Cambrian* newspaper, which some twenty-eight years ago had reported on the first locomotive journey, returned to cover the event:

'Locomotive Engine – On Friday last a new Locomotive Engine, called the Eclipse, started from the Penydarren ironworks, Merthyr Tydfil, with a load of 23 tons of bar and rod iron, which was delivered at the basin of the Glamorganshire Canal, 10 miles from the works, in one hour and forty-eight minutes. The engine then returned with the empty carriages, crowded with passengers, whom the novelty of the occasion had attracted, and reached the works, in one hour and 45 minutes, from the time of its leaving the basin, including all stoppages on the road for water &c. This was performed twice in the course of the day, and had there been occasion, it could have made another trip with the greatest ease. The day following it came up from the same point, drawing the same number of carriages, and conveying 145 passengers, in one hour and fourteen minutes, including stoppages; and as preparations are making to obviate the necessity of taking water on the road, it is expected that the journey will be accomplished in future easily and safely in an hour. This engine was built at Newcastle-upon-Tyne by Messrs R Stephenson, the celebrated Locomotive Engine-builders who have made these engines now travelling on the Liverpool and Manchester Railway; it has been much admired for the symmetry of its proportions and being very low and compact in general appearance, conveys to the mind a pleasing idea of smugness and aptitude for the task assigned to it...It has two 7 inch steam cylinders of 20 inches stroke, is mounted on four wheels 3 feet 4 inches in diameter and weighs when the boiler is charged with water, and in working trim, about five tons.'[37]

At Penydarren, a locomotive had returned to the tramroad, and if the name was intended to be symbolic of the return to the tramroad to 'eclipse' the original journey, then it was a successful one. While it did not result in the Penydarren company ordering any further engines, there was interest from the neighbouring Dowlais works. In August 1831, Dowlais were interested in ordering an engine

The Rocket *as an exhibit at the Patent Museum (now the Science Museum, London). The locomotive is seen in the condition in which it was presented to the Patent Museum. (SKJ collection)*

from the Neath Abbey ironworks, managed by a Quaker, John Tregelles Price (1783-1854). Another partner, his brother Henry Habberley Price (1794-1839), was well versed in civil engineering, having worked under Telford, and who was, in 1832, along with William Brunton (1777-1851), proposing a line between London and Bristol.[38] Price was to later support Brunel's routegave evidence against it.[39] Under John Tregelles Price's management, virtually every type of steam engine was built at Neath Abbey, including high-pressure steam engines on Trevithick's principles. The first of these is believed to have been supplied to the Swansea pottery in 1806.[40] From 1829 Neath Abbey began to manufacture locomotives although the Dowlais order was to be no ordinary locomotive engine, but one capable of rack or adhesion drive, the former enabling it to work the inclines. It was a novel approach, although rack locomotives had first been introduced by Blenkinsop and Murray, following Trevithick's work, as a way of reducing engine weight while improving traction. Here the rack was intended to be an auxiliary drive for steep gradients. Like Forman's engine, there were to be several changes along the way and, although Dowlais did receive an engine capable of adhesion and rack-drive, it was to go through an extensive process for which the Neath Abbey designers produced, once the general arrangement had been agreed, a series of fifty component drawings.

Among other changes, M.J.T. Lewis records that the original name of the engine should have been *Success,* but before it was finished the name was crossed out on several drawings, with the name Perseverance replacing it.[41] 'Perseverance' was not an uncommon motto in Wales; it was the Crawshay's family motto, which had also been passed on to the Glamorganshire Canal.[42] It was also the motto of the Sirhowy Tramroad, in which Harford, Davies & Co. of Ebbw Vale had a large stake. This company had ordered a Neath Abbey engine, and there is further confusion because the drawings for this engine, Industry, were also altered in the same way. Both engines were 0-6-0s, with ten 1½in cylinders, lowering chimneys and the same gauge (an earlier and unconnected Perseverance, built by a Mr Burstall, had been entered for the Rainhill Trials but was withdrawn when it met with an accident on route to Liverpool). There does not appear to be a good reason for giving a Dowlais engine the name Perseverance but, whatever the reason, Perseverance stuck, and the engine was delivered in June 1832. The total cost, in terms of the engine and mechanical equipment, was £762 8*s* 7*d*, which, although expensive compared to the cost of Eclipse, was for the first combined adhesion and rack locomotive which was capable of hauling iron all the way from Dowlais to Abercynon and back. Lady Charlotte Guest refers to this locomotive when travelling on the Liverpool & Manchester Railway in November 1833: 'The Engine is twice as fast, and many times more easy, than my little Locomotive... '[43]

The Cambrian covered the story, countering a misleading report from the *Hereford Times* that referred to the locomotive as the *Powerful*, introducing the story under the heading 'Progress of Science – Locomotive Engines', from a scientific correspondent:

> 'Some twenty years since Captain R Trevithick under the auspices of the late Mr S Homfray of Pennydarren [sic], attempted to convey iron from Pennydarren works to the basin on the Cardiff Canal, by steam power. The attempt not succeeding, horses have been constantly employed for that purpose since that time until the Eclipse from the Newcastle Manufactory of Messrs Stephenson, lately reported in your columns, accomplished successfully the desired effect – The Eclipse moves a small load of twenty or twenty-five tons, at the rate of eight to ten miles an hour. We are now credibly informed that the Eclipse is eclipsed by the Perseverance, got up by the Neath Abbey Iron Co., and supplied with the assistance of a Rack running parallel with the tram plates from Pennydarren to the Dowlais works. This engine has accomplished the amazing task of conveying from the Dowlais works to the basin aforesaid 126¼ tons of iron, besides the weight of engine, tender and wagons 50¾ tons making an aggregate of 177 tons. The engine after waiting several hours for the discharge of the iron returned with her complement of waggons and ascended the side of the mountain, by means of the Rack, with ease, without stopping for steam… '[44]

Dowlais were satisfied with the results, and another five engines of different types were to follow over the next six or seven years. Locomotives had returned to Merthyr and the Merthyr Tramroad, and proved the viability of steam power, but the tramroad would have to be replaced by the edge railway before all the elements of the modern locomotive railway were in place.

Chapter 4 Notes

1 Tomos, Dafydd. (1972). pp. 25-26. *Michael Faraday in Wales including Faraday's Journal of his Tour through Wales in 1819*. Denbigh: Gwasg Gee. The chapter title comes from Alfred Tennyson's poem 'Locksley Hall', see Warburg, Jeremy. (1958). The Industrial Muse. p.23. London: Oxford University Press, 'When I went by the first train from Liverpool to Manchester (1830), I thought that the wheels ran in a groove. It was a black night and there was such a vast crowd around the train at the station that we could not see the wheels. Then I made this line.' Locksley Hall was published in 1842, the line was, 'Let the great world spin for ever down the ringing grooves of change.'

2 Bye, Shiela. (2003). Early Railways Conference 2. p.142.

3 Lewis, M.J.T., (ed.). (2003). *Early Railways 2*, see Reynolds, Paul. *George Stephenson's 1819 Llansamlet Locomotive*. London: The Newcomen Society. Paul Reynolds has also researched the history of one of the later and most technically advanced tramroads in Wales: the Brecon Forest Tramroad. Built in 1821-1825, it represented advanced standards of engineering as far as the horse-drawn tramroad was concerned. Paul Reynolds (1979), *The Brecon Forest Tramroad*. Swansea: P.R. Reynolds. See also Stephen Hughes (1990). *The Archaeology of an early railway system: the Brecon Forest Tramroads*. Aberystwyth: Royal Commission on Ancient and Historical Monuments in Wales.

4 Hoole, K., et al edited by Jack Simmons. (1975). pp.17-18. *Rail 150 the Stockton & Darlington Railway and what followed*. London: Eyre Methuen.

5 PRO. Rail 667/1099. *Stockton & Darlington Railway*, 21 August 1838. As well as being Thomas Meynell's agent, Jeremiah Cairns had also subscribed to the SDR in 1821 and had sought tenders from the Dowlais ironworks on the supply of rails when the line was being built.

6 Hoole, K. (1975). p.9. *The Stockton & Darlington Railway*. Newton Abbot: David & Charles.

7 Davies, Hunter. (1977). p.57-58. *George Stephenson*. London: Quartet Books. Bella Bathurst (2000). p.100. *The Lighthouse Stevensons*. London: Ted Smart. The latter refers to a proposition that they, Rennie and Stevenson, consider jointly surveying the route.

8 GRO. D/DG Letter Book, No.175, 18 May 1821. Robert Stevenson to George Overton.

9 GRO. D/DG Letter Book, No.177-79, 180-83, see also Elsas, Madeleine. (1960). *Iron in the Making*, the Dowlais Iron Company Letters 1782-1860, p. 60. Cardiff: Glamorgan County Council.

10 Davies, Hunter. (1977). *George Stephenson*. p.61. London: Quartet Books.

11 GRO. D/DG Letter Book, No.81, 1 August 1822, George Overton to J.J. Guest.

12 GRO. D/DG Letter Book, No.337,338, 12 June 1821, Jeremiah Cairns to J.J. Guest.

13 GRO. D/DG Letter Book, No.338 (copy letter), 9 June 1821, William Jessop to Jeremiah Cairns

14 Lewis, M.J.T., (ed.). (2003). *Early Railways 2*. Reynolds, Paul. *George Stephenson's 1819 Llansamlet Locomotive*. Reference to letter in Newcastle City Library [NCL], Tomlinson Collection on Early Railways 6749-53/1/5. (Letter 3).

15 GRO. D/DG Letter Book, No.338 (copy letter), 9 June 1821, William Jessop to Jeremiah Cairns.

16 Overton, G., (1825). *A Description of the faults or dykes of the mineral basin of South Wales: Part 1*. London.

17 GRO. D/DG Letter Book, No.279, 13 May 1821, Thomas Meynell to Dowlais Iron Company.

18 GRO. D/DG Letter Book, No.78, 30 July 1822, George Overton to J.J. Guest.
19 GRO. D/DG Letter Book, No.340, 1 November 1821, Jeremiah Cairns to J.J. Guest.
20 Carr, J.C. & Taplin, W. (1962). *History of the British Steel Industry*, p.7. Oxford: Basil Blackwell.
21 Trans. Newcomen Society. Vol.69, No.1, 1997-98, p.136.
22 Trans. Newcomen Society. Vol.69, No.1, 1997-98, p.136.
23 Todd, A.C. (1967). *Beyond the Blaze: A Biography of Davies Gilbert.* p.92. Truro: D. Bradford Barton. Quote from Giddy's (Gilbert) Diary 25 October 1809.
24 Tomos, Dafydd. (1972). p.25. *Michael Faraday in Wales including Faraday's Journal of his Tour through Wales in 1819.* Denbigh: Gwasg Gee.
25 Guest, Revel & John, Angela V. John. (1989). pp.124-25. *Lady Charlotte A Biography of the Nineteenth Century. London*: Weidenfeld and Nicolson.
26 Tomos, Dafydd. (1972). p.23. *Michael Faraday in Wales including Faraday's Journal of his Tour through Wales in 1819.* Denbigh: Gwasg Gee.
27 Tomos, Dafydd. (1972). pp.25-26. *Michael Faraday in Wales including Faraday's Journal of his Tour through Wales in 1819.* Denbigh: Gwasg Gee.
28 Gross, J. (ed.). (1976). *Merthyr Historian.* Vol. 1. Chapter II part 7, Gross, J. *The Merthyr Tramroad.* p.53. Merthyr: Merthyr Tydfil Historical Society.
29 Rattenbury, Gordon & Lewis, M.J.T. (2004).p.55. *Merthyr Tydfil Tramroads and their Locomotives.* Oxford: Railway and Canal Historical Society. M.J.T. Lewis's section was originally published as *Steam on the Penydarren*, Industrial Railway Record, No.59, 1975.
30 Rattenbury, Gordon & Lewis, M.J.T. (2004). p.59. *Merthyr Tydfil Tramroads and their Locomotives.* Oxford: Railway and Canal Historical Society.
31 Rattenbury, Gordon & Lewis, M.J.T. (2004). p.60. *Merthyr Tydfil Tramroads and their Locomotives.* Oxford: Railway and Canal Historical Society. The bill for the two engines working for William II in 1830 was submitted by Braithwaite.
32 Rattenbury, Gordon & Lewis, M.J.T. (2004). p.58. Merthyr Tydfil Tramroads and their Locomotives. Oxford: Railway and Canal Historical Society.
33 Rattenbury, Gordon & Lewis, M.J.T. (2004). p.58. *Merthyr Tydfil Tramroads and their Locomotives.* Oxford: Railway and Canal Historical Society and Mear, John F. (1999). p.66. *Aberdare, the Railways and the Tramroads.* Aberdare: published by the author.
34 *The Cambrian*, 18 March 1830. Letter to Cambrian by W. Crawshay Jr.
35 This was the refurbished Bryngwyn tramroad which the Crawshays had rebuilt and extended to Park Pit (Pwll Als) shortly before 1830, see Mear, John F. (1999). p.65. *Aberdare, the Railways and the Tramroads.* Aberdare: published by the author.
36 *The Cambrian.* Letter by William Crawshay sent to *The Cambrian*, 20 June 1829.
37 *The Cambrian*, 30 June 1832.
38 MacDermot, E.T., revised by Clinker, C.R.. (1964). *History of the Great Western Railway*, Vol 1. London: Ian Allen. The scheme for the 'Bristol and London Railway' was issued on 7 May 1832.
39 MacDermot, E.T., revised by Clinker, C.R. (1964). p.13. *History of the Great Western Railway*, Vol.1. London: Ian Allen.
40 Ince, Lawrence. (2001). pp.48-49. *Neath Abbey and the Industrial Revolution*, Stroud: Tempus Publishing Ltd.
41 Rattenbury, Gordon & Lewis, M.J.T. (2004). p.65. *Merthyr Tydfil Tramroads and their Locomotives.* Oxford: Railway and Canal Historical Society.
42 It was even used as a motto by the Victorian photographer, T. Forrest of Pontypridd.
43 Bessborough, Earl of, (ed.). (1950), p.19, *Lady Charlotte Guest Extracts from her Journal 1833-1852.* London: John Murray.
44 *The Cambrian*, 18 August 1832. A report of *Perseverance* working on the Merthyr Tramroad is somewhat misleadingly referred to as the *Powerful* in the *Hereford Times* for 4 August 1832, see Rattenbury, Gordon & Lewis, M.J.T. (2004). p.68.

5

Iron Link

'My First Child, My Darling'[1]

Brunel had started work for his father on the Thames Tunnel on 2 March 1825, a position that continued until work came to an end in 1828. The cessation of works was, in an indirect fashion, to lead to what he always regarded as his first independent engineering commission.

Following the flooding of the tunnel, Marc was concerned about the physical state of his son and that he should recuperate and recover from the injuries he had suffered. Brighton was the first destination where he appears, along with other diversions, to have taken the opportunity to take a close look at the famous chain pier: '... I met some pleasant company – strolled on the pier smoking my meerschaum before breakfast...'[2] The chain pier, designed by Capt. Sir Samuel Brown RN with four 255ft-spans suspended from cast-iron towers in an Egyptian style, was opened some five years previously in 1823.[3] It was the first pleasure pier, facilitating the embarking and landing of ships' passengers and offering promenaders a walk to a pierhead some 1,134ft into the open sea.

It was not only the pier that offered so many distractions as Isambard '... went to a Fancy Ball, etc etc' with the result that he suffered a relapse and had to be brought home. After recovering for some months, the traditional story refers to him being sent to Clifton, near Bristol, where there would be little temptation for the 'etceteras' of life, and where he could relax and sketch the scenery but, no doubt, also exert himself climbing in the gorge. The poet George Crabbe (1754-1832) wrote of the recuperative affects of the location in 1831: 'Clifton was always a favourite place with me. I have more strength and more spirits since my arrival at this place...'[4] This period of rest and recuperation is not recorded by Brunel in his journals, although there is a reference, in the summer of 1828, of him going for a '... trip to Plymouth, where he examined with great interest the Breakwater and other engineering works in the neighbourhood'.[5] There is no specific entry for Bristol until, possibly, the following year.[6] Nevertheless, he appears to have knowledge of the Clifton area: '... in the autumn of 1829, he heard that designs were required for a suspension bridge over the Avon at Bristol, and he determined to compete.'[7]

The story of the bridge competition really begins in 1754, with the death of William Vick, a wealthy Bristol wine merchant who left £1,000 in his will. It was left for a specific purpose; that it should be invested until it reached £10,000, a sum which would be sufficient (he was assured) to build a stone bridge, free of toll, across the Avon Gorge. It is not entirely clear why Vick should want to build a bridge here as at the time of his death Clifton was still a hamlet and Leigh Woods, the other side of the gorge, was dominated by private estates. The Georgian Clifton of grand terraces was yet to be developed. There was no doubt it would be a bold statement in this setting, crossing from one side of the gorge to the other, and that may have been reason enough.

William Bridges was the first to give Vick's legacy a potential outcome in the form of a multi-purpose and multi-storey bridge design in 1793. His spectacular stone bridge was a miniature town in itself and included a museum, market, library, offices, houses, a tavern, stables, corn exchange, coal store, granaries and wharves. Even this was not enough as a lighthouse, two windmills and a chapel with belfry and weather vane were to be built above the great arch. Vick's money was still accruing interest but it was clearly obvious that the cost of William Bridges' design would have been prohibitive.

A cheaper alternative to stone was iron and, from the turn of the nineteenth century, suspension bridges of wrought iron were being built in the United States of America by James Finley. This type of bridge was to be pioneered in Britain by Capt. Sir Samuel Brown RN, who collaborated with Thomas Telford on a joint proposal to cross the Mersey at Runcorn in 1816. It was never built but Brown was to patent his design of suspension chains, consisting of flat eyebar links, in 1817.[8] Telford was also to use this design on his suspension bridges at Menai Bridge and Conway. Brunel had taken the opportunity to view the Menai Bridge shortly after it opened in 1826, following a meeting on his father's behalf for the proposed Grosvenor Bridge in Chester, and was to re-examine this bridge and others over the Tyne and Tees in preparation. Brown's first large suspension bridge was the 1820 Union Suspension Bridge over the Tweed near Berwick-on-Tweed. It still stands today, crossing from England to Scotland, with a span of 361ft.

With Vick's proposed sum of £10,000 for the bridge becoming a realistic goal when the fund reached £8,000 in 1829, a committee was set up to decide how best to build a bridge. It was soon realised that a stone bridge could not be built for less than £90,000. It was more realistic to bridge the gorge with an iron suspension bridge, but that would still require tolls to finance it, even though that was against Vick's original legacy. An Act of Parliament was passed to allow these changes and, on 1 October 1829, a competition was announced that specifically called for: '... DESIGNS for the ERECTION of AN IRON SUSPENSION BRIDGE at CLIFTON DOWN over the River AVON...'[9]

Brunel was just twenty-three, and was to submit four plans enlisting the expertise of his father, with spans ranging from 760ft to 1,180ft between points of suspension, but at first his entry seemed to have little chance. His favourite design involved a tunnel which emerged from the Clifton side of the gorge onto the bridge, and

Brighton Chain Pier. Opened in November 1823, the chain links were supplied by Capt. Sir Samuel Brown's Newbridge chainworks. (SKJ collection)

through another tunnel on the opposite side. In all, twenty-two plans were submitted for the competition, including a design by Capt. Sir Samuel Brown RN, who was the most experienced builder of suspension bridges in the country. His collaborator on the earlier Runcorn Suspension Bridge, Thomas Telford, was to be brought in by the committee to judge the entries.

Telford was Britain's leading civil engineer but, now in his seventies, he was more cautious in his outlook, particularly with regard to the maximum span of such a bridge. He stated that 600ft was the longest possible span and summarily dismissed all the entries. The committee then had no other recourse than to ask for his proposal, and he produced a plan for a three-span bridge supported by two Gothic towers rising from the banks of the river, with a centre span of 360ft. Although presented to the public with great ceremony, it did not captivate the public for long and eventually aroused dissatisfaction and ridicule. With public opinion against Telford's design, the committee shelved it on the grounds that it would be too expensive to build, and the following year announced a second competition, in which Telford was allowed to compete but not to judge. The role of referee was filled by the president of the Royal Society, who was none other than Davies Gilbert (Giddy), the friend and mentor of Richard Trevithick.

Gilbert knew something about suspension bridges and indeed, had been in communication with Telford during the construction of the Menai Suspension Bridge. This stemmed from a long interest in the catenary curve, which dated back

Clifton Suspension Bridge. (SKJ photograph)

to 1791, and related to Tom Paine's bridge: 'The bridge was an inverted Catenary, and the attention I then paid to the nature of the Catenary curve was, I believe, the means of my afterwards writing on the Menai Bridge in the Philosophical Transactions.'[10] As a commissioner of the Parliamentary Roads to Ireland, he had a professional interest in the Menai Bridge and was to expound his mathematical calculations concerning the form of the curve to Telford. He wrote to journals, such as the London Quarterly Journal of Science, to outline his theorems on the subject, with mathematical values relating to Menai, and offered a simple conclusion on the strength and safety of suspended bridges: '… their points of attachment cannot be too lofty, nor consequently the curvature of the chains too great.'[11] Consequently, Gilbert was to record in his diary: 'By my recommendation the Towers were raised and the curvature of the chains increased, so as to nearly double the strength, the whole at the expense of 15 Thousand Pounds.'[12] Gilbert was therefore well versed in suspension bridge theory, but was not a practical engineer. To compensate for this he was assisted by John Seaward (1786-1858) who, as well as being a mathematician who had published a paper on suspension bridge chains, was a partner in a foundry and built marine engines.

Twelve designs were submitted for the second bridge competition, and five were retained for the attention of Gilbert and Seaward. They announced their report on 16 March 1831. Brunel's 630ft-span design was one of four to be shortlisted; Telford's design was the first to be rejected. Brunel's design (he had again submitted four entries) was placed second, with that of W. Hawks first. Brunel was critical of Seaward's opinion, and believed he was responsible for his placing in the competition, based on his evaluation of aspects of the chain and suspension-rod design, and details of the anchorage. That Brunel was already known to Gilbert is clear from journal entries in which Brunel records making drawings of geological data for Gilbert in February 1829.[13] Brunel quickly arranged a meeting with Gilbert, and convinced him that the technical objections to his design were unjustified: 'D.G. having recanted all he had said yesterday I was formally appointed and congratulated very warmly by everybody.'[14] As well as his design being chosen as the winner, Brunel was also appointed engineer for the project, and his favourite 'Egyptian' design was adopted. What the original winner of the competition thought of the judges' change of heart is not recorded, but Brunel was to record in a letter to his brother-in-law Benjamin Hawes, dated 27 March 1831:

> '… I think yesterday I performed the most wonderful. I produced unanimity among fifteen men who were all quarrelling about the most ticklish subject – taste. The Egyptian thing I brought down was quite extravagantly admired by all and unanimously adopted; and I am directed to make such drawings, lithographs, etc as I, in my supreme judgement, may deem fit; indeed they were not only very liberal with their money, but inclined to save themselves much trouble by placing very complete reliance on me.'[15]

His friend, the artist John Callcott Horsley (1817-1903), later to become his brother-in-law when Brunel married Mary Horsley in 1836, elaborated on this Egyptian style to which Brunel had entrusted him with a special commission:

> 'As originally designed it was to have cast-iron towers, purely Egyptian in form, and decorated on the panels into which the faces were divided with incised figure designs, illustrating all the processes necessary in the production of the various portions of the bridge. He had himself made spirited outline sketches in pen and ink of a few of the leading subjects…' [16]

It appears that Brunel drafted these 'spirited outline sketches' soon after receiving the Clifton commission and presumably when he was visiting and seeking potential suppliers of the bridge ironwork from various ironworks, such as those at Merthyr Tydfil. The earliest reference to Horsley travelling to Wales appears to be in November 1834, according to his sister, Fanny Horsley; '… you may conceive John's delight at seeing Wales and the largest ironworks in the (country).'[17] That he made these sketches in line with commissioning further drawings, lithographs and so on as 'may deem fit' is further qualified by Horsley:

> 'He made very clever sketches for some of these proposed figure subjects, just to show what he intended by them. I remember a group of men carrying one of the links of the chainwork, which was excellent in character. He proposed that I should go down with him to Merthyr Tydfil, and make sketches of the iron processes. We accomplished our journey, and all the requisite drawings for the intended designs were made.'[18]

If Brunel had made these sketches from life, it is possible that the chain links were sketched at Brown Lenox at Pontypridd, which was the company that Brunel was particularly keen should tender for the chainwork. However, the visits that Horsley was involved with were made when Brunel was engaged on the surveys for the Taff Vale Railway so the 'very clever sketches' he made of the proposed figure subjects that were to decorate the cast-iron casing of the towers are likely to have been carried out before he was actively engaged on TVR business. Horsley refers to Brunel's conception of the towers or gateways at either end of the bridge to be peculiarly grand and effective: 'They were to be purely Egyptian; and, in his design, he had caught the true spirit of the great remains at Philae and Thebes.'[19] He describes how the subjects would have been arranged in tiers from the top to bottom of the towers, illustrating the whole work of constructing the bridge, from quarrying the iron ore, making the iron and the fabrication of the individual ironwork elements of the bridge, to the construction of the towers and the final completion. Horsley was well qualified to undertake the drawing work required; he was to become a member of the Royal Academy and in 1840 designed the first Christmas greetings card for his friend, Sir Henry Cole.

While the 'Egyptian thing' had been going in and out of style for centuries, it had undergone a major revival following interest in Napoleon's Egyptian campaign of 1798. This was to fuel a demand for all things Egyptian – decoration, furnishings and antiquities. Industrial and engineering structures also received the Egyptian treatment; Capt. Brown incorporated the style into his Brighton chain pier of 1823, and an Egyptian temple was the inspiration for the Leeds flax mill of 1842, which still stands today. An earlier industrial group that received the Egyptian treatment was the Bute ironworks in the Rhymney Valley, which featured three blast furnaces based on the architectural ruins of Dendyra. Here, acting on behalf of the second Marquess of Bute, the surveyor, David Stewart, persuaded William Forman (*d.*1829), then the principal Penydarren partner, to erect a new ironworks. Work on what began as an ambition of the Marquis of Bute, to build an ironworks that would rival the Merthyr ironworks, particularly Dowlais, began in 1824. The Bute ironworks, Stewart claimed, '… may become the first works in Wales and consequently in the world.'[20] The works were therefore more ostentatious than most, having three 'pylons' – one for each of the blast furnaces – which were linked together at high level by arches with a continuous frieze at the top.[21] What also stood the Bute ironworks apart from other Egyptian-style buildings was that one of the works buildings was modelled after a monument to the Newtonian Ball, also in the Egyptian style, and which was believed to be derived from the designs of Etienne-Louis Boullee (1728-99) for a tomb of Newton.[22] The structures were given much publicity through articles and comments that appeared in the engineering press, and a series of drawings and engravings exhibited at the Royal Academy in London in 1828. Whether this example was an influence is difficult to say, although Brunel was believed to have based his Egyptian gateways on the examples found in the ruins of Tentyra, probably the Temple of Hathor at Dendyra.[23]

The *Bristol Mercury* confirmed that:

> 'The design chosen by the committee is that of Mr. Brunel jun...' and that the sum estimated to complete the erection of the Clifton Suspension Bridge, including the approaches and every expense amounts '... to £57,000 of which about £20,000 yet remains to be provided.'[24]

The *Bristol Mercury* was reporting on a somewhat premature ceremony that took place to mark the formal start of work on the morning of 21 June 1831. Speeches were made by Lady Elton and Sir Abraham Elton during the stone-laying ceremony. In his speech, Sir Elton alluded to Brunel being remembered by future generations as: '... the man who reared that stupendous work, the ornament of Bristol and the wonder of the age.'[25]

The age was a troubled one, however, with Britain in the grip of agitation over reform measures. The first Reform Bill had been thrown out by the House of Lords in March 1831, and the rejection of the Second Reform Bill in October 1831 was to precipitate some of the bloodiest civil disturbances ever to take place in nineteenth-century England. George Crabbe was still in Clifton on 31 October 1831, and wrote:

> 'Queen's Square is but half standing; half is a smoking ruin. The Mayor's house has been destroyed, the Bishop's palace plundered, but whether burnt or not I do not know. This morning a party of soldiers attacked the crowd in the Square; some lives were lost, and the mob dispersed, whether to meet again is doubtful. The military are now in considerable force, and many men are sworn in as constables...'[26]

Brunel was one of those sworn in and was in the thick of the action, protecting and rescuing treasures from the Mansion House. In this episode he was to meet up with Nicholas Roch (*d.*1866), a prominent Bristol businessman from an old Pembrokeshire family who was to play a crucial role in Brunel's future career. After three days of mob rule in Bristol, order was eventually restored by military force, but all hope of a quick resumption of work on the bridge was dashed.

Bristol was not alone in terms of political disturbances; Merthyr was gripped by radical fervour following economic recession in the summer of 1831. After a month of increasing disorder, riot and insurrection by the working population of Merthyr, erupted on 2 June 1831. The major event of this disturbance was centred on the 'Castle Inn', where the High Sheriff of Glamorgan was waiting with the ironmasters and magistrates. Troops opened fire and, after a protracted struggle in which twenty protestors were killed and a hundred injured, the street was eventually cleared. Of all the rioters that were subsequently indicted, one, Richard Lewis (alias Dic Penderyn) was to be hanged at Cardiff for his role in the disturbances. There was controversy over the evidence used to secure his sentence and many at the time protested against the harsh sentence, including the Quaker ironmaster, John Tregelles Price (whose Neath Abbey works was to supply the Dowlais locomotive engine).

William Crawshay II and John Guest were also present at the 'Castle Inn' and appealed to the crowds to disperse amid cries of 'Bara neu waed' (bread or blood). Today, Dic Penderyn is revered as a working-class martyr.

The Clifton Bridge Committee were about to launch a fresh appeal for funds in October 1831, but with the nationwide unrest, in which Bristol had a particularly high profile, business confidence was low and raising funds for the bridge was deemed a lost cause in the short term. Over four years would elapse before work on the bridge resumed. The encasement of the towers was never carried out; Brunel was to make a number of cost-saving measures which included abandoning the Egyptian ornamentation in 1835. Neither Brunel's sketches nor the drawings undertaken by Horsley for this work have come to light.[27] An example of what these relief clad towers may have looked like can today be found in St Petersburg, where two pylons, designed by the Scottish architect, Adam Menelas (1753-1831), form the entrance to Alexandra Park at Tsarkoye Selo. The pylons are brick-built and faced with cast-iron plate reliefs. Work on the project by Menelas for Tsar Nicholas I had begun in 1827.[28] They were completed in 1830.

Clifton had been a high spot in a phase of Brunel's career that was otherwise to be marked by frustration. It had started optimistically enough with the Clifton Bridge, but even that was to turn into an uncompleted dream. Setbacks included the failure of the 'Gaz' engine experiments, a project Brunel started with his father and had pursued for ten years. Several minor commissions were carried out by Brunel in this period, such as drainage works at Tollesbury on the Essex coast. A more important commission materialised in November 1831, for the construction of a new dock at Monkwearmouth near Sunderland, but Parliament rejected the first designs and, when an Act was eventually passed, it was for a scaled-down scheme. Another dock survey, this time for the Navy Yard at Woolwich, came to nothing. One commission that was carried out in this period was an observatory with a revolving dome which was constructed by Maudslays, for Sir James South. It was completed by May 1831 but, because the original estimate had been exceeded, South refused to pay. To justify his reasons for non-payment, he publicly criticised the design, and it was an episode that deeply hurt and incensed Brunel.

In August 1832 a more positive dock commission was in the offing. This was for Bristol, and had come through a contact he had made with a fellow public-spirited individual when they were protecting and rescuing treasures from the Mansion House during the riots. This was Nicholas Roch, who introduced him to the directors of the Bristol Dock Co. Bristol's position as one of the first rank of ports in the country was suffering from the state of the Floating Harbour and the build-up of silt, which had stubbornly failed to be efficiently scoured away by various improvements. From the 1820s the problem was compounded by the influx of sewage and the shoals continued to accumulate presenting a constant hazard and impediment to shipping. Although Brunel was commissioned to report on improvements to the dock, there was reluctance and a lack of confidence to invest in new works following the Bristol riots, but eventually the dock committee was to agree to undertake work under Brunel's supervision in 1833-34. Only one of his major proposals, the conversion of the Rownham Dam from an 'overfall' weir to an 'underfall', was started, along with the introduction of a purpose built 'drag-boat'. This drag-boat remained in working order at the docks until the 1960s[29] and a near-contemporary example, also built by Brunel but for Bridgwater Docks, is claimed to be the oldest surviving iron boat.[30]

There were other opportunities on the horizon for the engineer in this period, in the form of railways. Locomotives had even been successfully reintroduced to the Merthyr Tramroad and numerous proposals for new routes were put forward following the success of the Liverpool and Manchester Railway in 1830. Brunel had been trying to secure a commission for one of the new railway schemes for some time but his attempts were largely unsuccessful as he had little practical experience, and he faced established preferences for the Stephenson school of railway engineering. One story relates an occasion when Brunel was in the north of England and George Stephenson took him playfully by the collar, saying: 'My good friend, understand that you are not to appear north of the Trent!' Despite this, Brunel was to become a close friend of Robert Stephenson and, in spite of public disagreements over professional matters, as the leading engineers of their day they would support each other in times of crisis.

In March 1833, the merchants and entrepreneurs of Bristol decided it was time to respond to the challenges of a new era by investing in railway communications. With the contacts made through the Clifton Suspension Bridge and Bristol docks, Brunel was, at the age of twenty-seven, to be appointed engineer to a proposed railway connecting Bristol with London. A line was to become famous as the GWR, the story of which is too well known to repeat here except for a brief outline to compare with a subsequent railway commission. In Brunel's mind the GWR was not to be like any other railway that had been built – not a tramroad partly operated by steam locomotives but a line of communications, built from scratch. Brunel visualised the GWR as the start of a growing network that would not be confined solely to London and Bristol. In order to construct a railway that would be a major improvement on any previous one, the principles of the construction and working had to be completely rethought. The rate of travelling could be greatly increased without risk to the public, and it was this philosophy that led Brunel to build on a grand scale. The scale and proportions of his railway were to exceed those constructed or designed by the Stephensons after the model of the waggonways of the north of England.

For the GWR, Brunel adopted a gauge of 7ft which was to be a characteristic feature of all his British railways, with the sole exception of the Taff Vale Railway. The broad gauge, as it came to be known, offered advantages such as an improved locomotive layout giving a lower centre of gravity. Although these advantages were never fully exploited, the broad gauge represented a challenge of speed to the narrow-gauge railways and was a general stimulus to high-speed railway operation. Brunel referred to the GWR as '... the finest work in England' and was to build it accordingly. Part of the development of a rail track suitable for this involved a new design of rail for which John Guest and the Dowlais works were to play a prominent role. On behalf of Brunel, George Frere had written to Thomas Evans of the Dowlais works in

October 1836, taking up an offer by Guest, '... to further the improvements of railways by rolling rails for the purpose of experiment.'[31] Guest was to develop a unique rail section that could accommodate both tram wheels and flanged wheels as part of the changeover from tramroads to edge rails that was now taking place. This rail had an inverted 'U' profile from which a flat plateway section extended on one side. It was to be rolled in great quantities at Dowlais and became known as the 'Guest' rail. It appears to have been a development of the design rolled for Brunel which became known as bridge rail or more commonly as 'Brunel' rail. Lady Charlotte Guest was to record in her journal in November 1836 about the experiments taking place: '... after dinner I went with Merthyr [her pet name for Guest] to the Upper Mill to see them try the rolls for a new sort of rail which Mr Brunel talks of laying down on the Great Western Road.'[32] Brunel's ideas on this can be seen in his sketch book, starting as a solid rail section with flanges or feet extending either side to spread the load on the longitudinal timbers he was proposing to use on his broad-gauge road.[33] Brunel appears to have been pleased with the results and, early in 1837, George Thomas Clark was to write on his behalf to the Dowlais Iron Co.: '... Mr Brunel wishes you to roll 25 tons – of which quantity 20 tons are to be sent to London and 5 tons to Bristol...' [34]

In terms of the standard of engineering works to be designed and built for the GWR, Brunel had obviously learnt a great deal from his father and his experience with the Grosvenor Bridge may also have been an influence. The first section of the GWR to be completed crossed the Thames at Maidenhead, where Brunel designed a bridge with two semi-elliptical arches of 128ft and a rise of only 24ft 3in. These were the flattest and possibly the largest arches ever constructed in brick work.

The last section of the GWR to be completed included the formidable Box Tunnel. At nearly 2 miles long, it was by far the longest railway tunnel attempted at that time. Heavy works, such as cuttings and embankments, were necessary for the gradients Brunel required. These gave the GWR the nickname of 'Brunel's billiard table'. The easy gradients and gentle curves on the line, together with the advantages offered by the broad gauge, enabled the GWR to operate the fastest trains in the world. The year 1841 saw the completion and opening throughout of the GWR between London and Bristol. Paralleling the development of the GWR was the construction of his only 'narrow' gauge line, the Taff Vale Railway, which was also opened throughout in 1841.

Work on Brunel's 'first child' was not completely forgotten. In 1836 an iron bar of 1¼in-diameter was hauled across the Clifton Gorge, from which was slung a basket to transport men and materials across. Brunel himself made the first crossing but when the basket stuck at the lowest point of the bar, some 200ft above the river, he climbed onto the edge of the basket to release the cable from the jammed pulley. That this mode of crossing was, to say the least, highly dangerous, is confirmed by another intrepid traveller, Thomas Sopwith (1803-79). In June 1840 he recorded in his notebook: 'I crossed the River Avon on the iron bar from pier to pier and narrowly escaped an accident – the prospect was truly sublime.'[35] The foundation stone of the bridge was laid in 1836, coinciding with the British Association for the Advancement of Science meeting which took place in Bristol at the end of August.

In July 1839 advertisements were taken out for the supply of ironwork for Clifton Suspension Bridge, giving the deadline as 14 September 1839.[36] The following month, William Purnell of the Dowlais Iron Co. wrote to Thomas Evans, the London agent, to comment on hearing that Brunel had written to Brown Lenox, telling them that '14th proximo' was the deadline for sending in a tender for the Clifton Suspension Bridge. Purnell felt that: 'Due to their prior experience, Brunel will give them preference.'[37] Dowlais were reluctant to become involved, even as a supplier of bar iron, and initially decided to favour two of the potential chain link contractors, Brown Lenox (Brunel's preferred contractor) and Sandys, Carne & Co., by quoting the same price per ton of £11 10*s* 0*d* to £12 per ton (Acraman & Co. of Bristol had been quoted £12 10*s* 0*d* per ton for the bar iron). However, ten days later, the company decided that they would not tender for Clifton Bridge iron at less than £12 10*s* 0*d* per ton as it was 'a troublesome order'.[38] Presumably, Brunel was persuasive as not only did Dowlais supply the iron but at a lower price! John Guest was told on 21 September 1839 that Brunel wished to give Dowlais the order for the Clifton Bridge, his ideal combination being that: '... Brown Lenox did the smith's work.'[39] However, with no Brown Lenox tender being submitted for the work, Brunel chose Sandys, Carne & Co. (or Sandys, Carne & Vivian of the Copperhouse Foundry, Hayle, as they were also known) and reported in August 1840 that a satisfactory contract had been entered into.[40]

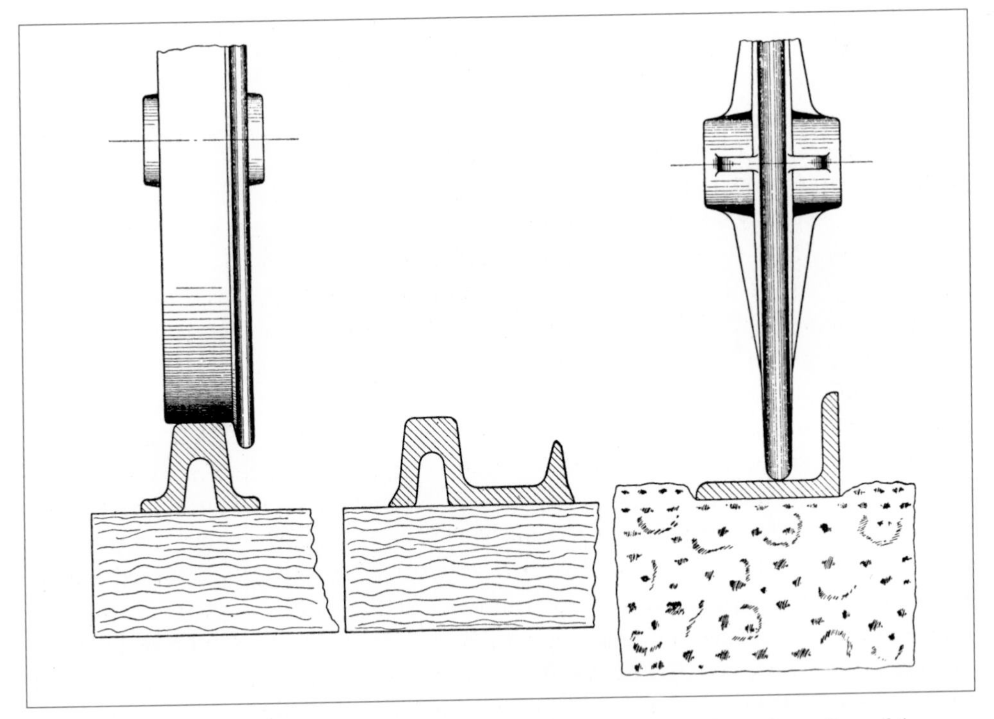

From right to left: the transition from plate rail to Brunel rail via Guest rail. (From Proceedings of the South Wales Institute of Engineers, Vol.III, 1937)

The Copperhouse Foundry had sought the prices of bar iron from Dowlais in a letter to Thomas Evans of 15 August 1839, and in November 1840 wrote again enquiring when they would have ready: '... from 80 to 100 Tons of the Bars for the Clifton Bridge at Cardiff, as We shall order our Vessels accordingly...'[41] Dowlais had finally agreed only to roll the bars and to supply the 7in-wide by 1in-thick bar iron in various lengths at £10 10*s* 0*d* per ton by June 1840.[42] Copperhouse would be responsible for the conversion of the iron into chain links with the welding of eye plates on each end. John Pool of the Copperhouse Foundry was to make several complaints about the standard of bars due to the rejection by Brunel's agent, Charles Gainsford, of a number of the supplied bars. Despite this, Sandys, Carne & Vivian returned to Dowlais when they tendered for another of Brunel's suspension bridges, the Hungerford Suspension Bridge. In 1843 they requested that the 7in-bar, eye plates and small iron was: '... of the same quality of iron made by you for the Clifton Suspension Bridge.'[43]

At Hungerford the architect James Bunning (1802-63) was also involved in the design for a footbridge across the Thames at London. It was completed in 1845 with a main span of 676ft. As with Clifton, Dowlais was to supply the 7in-bars onto which

Example of Guest rail at Merthyr. (SKJ photograph)

the eye plates would be welded by the Copperhouse Foundry. Hungerford Suspension Bridge was to be demolished before Clifton was completed. The salvaging of the Hungerford chains led to the completion of Clifton.

At Clifton, Brunel was asked to stop work on the chain iron in December 1842. By 1843 the two towers were completed but funds were exhausted despite abandoning the cast-iron cladding with its Egyptian ornamentation topped with sphinxes, and a reduction in height of the towers. By 1853 the time span allotted by Parliament had expired and the committee sold the ironwork, machinery and equipment to pay the contractors. Brunel ensured that some of this material was used on his Royal Albert Bridge carrying the broad gauge across the river Saltash between Devon and Cornwall. A total of 2,800 or so chain links were used as tension members on the Royal Albert Bridge, some 1,150 of these being the chain links originally supplied for the Clifton Bridge using Dowlais iron.

Brunel's early death inspired the completion of the bridge and it was eventually finished in 1864, five years after his death, in part as a memorial by his fellow engineers, led by the Institution of Civil Engineers. Hungerford was about to be demolished to build a railway bridge, and the chains were bought for £5,000. A new bridge company was formed in May 1860, and it was estimated that £45,000 would be needed to complete the project, £30,000 of which had been raised by December. A new Act of Parliament was obtained in 1861 to complete the bridge under the supervision of Sir John Hawkshaw (1811-91) and William Henry Barlow (1812-1902). There were a number of changes to Brunel's design, some of which reflected advances in suspension bridge design, namely the bridge deck which was to be stiffened by a wrought-iron lattice work girder, which replaced a timber deck. Brunel's last proposal in this area was for a bridge deck with two longitudinal timber beams over 6ft deep.[44] The strength of the suspension cable was also increased by an additional chain on each side i.e. from two to three. This was fortunate for today's road traffic, as was the widening of the carriageway to 20ft wide.

Barlow records that work began as soon as: '... the chains of the Hungerford Bridge were set at liberty...'[45] Hawkshaw was the engineer for the Charing Cross Railway Bridge being constructed on the Hungerford site. With the chains secured for Bristol by the beginning of 1863, timber scaffolding was placed around the towers and an iron wire was fixed across the Clifton Gorge. From this, a temporary suspended staging and walkway was built up, which enabled the chain links to be assembled at the rate of forty a day and, when the three chains on the south side were completed, the walkway was dismantled and re-rigged for the northern side. A total of 4,200 bar links were assembled and by May 1864 the chains were complete. Wrought-iron suspension rods, 162 in total, were fixed to the chains and supported the two large wrought-iron lattice work girders that ran the full length of the bridge. Cross girders were then bolted underneath at right angles to make the structure rigid, and the timber road deck was made up. Prior to opening, a dead weight of 500 tons of stone was distributed over the surface, causing a total deflection of 7in in the centre of the bridge, which was well within acceptable limits.

The bridge was ready for public traffic on 8 December 1864 and the occasion was marked by a major public ceremony to mark the final realisation of William Vick's vision. The span of the bridge was now 702ft 3in, with a clear headway of 248ft above high water, and a deck width of 31ft including roadways and footways. The contractors for the bridge were Cochrane Grove & Co. of Dudley, who also dismantled the Hungerford Suspension Bridge and built the bridge that replaced it, Charing Cross Railway Bridge. Hungerford was to supply 1,040 tons of ironwork, and some 200 tons of new links were required for the additional third link on each side. Thus, some ⅔ of the chain that supports the Clifton Suspension Bridge today was part of the original order of Dowlais bar iron to Hungerford, the 'new' links being supplied by Cochrane Grove from the Lord Ward Round Oak works in Dudley.[46]

Brunel had started this period in a relatively unknown capacity, albeit with a famous father, but had now become an engineer with a growing reputation. The Bristol connection and the three commissions he gained there – Clifton Suspension Bridge, Bristol docks and the GWR – were of great importance in this, the latter commission being the most important in the long term but one he may not have achieved without the progression through the first two. Brunel, writing on the

Box Tunnel, seen from an escorted BR visit immediately after dawn on Brunel's birthday, 9 April 1973. Even Sir Allan Quartermaine, the last chief civil engineer of the GWR and the first chief civil engineer of the Western Region of British Railways, was not aware of the story of the sun shining through Box Tunnel on 9 April.[47] (SKJ photograph)

When Brunel married Mary Horsley on 5 July 1836, he managed to take time out of his busy schedule to honeymoon in North Wales. The Capel Curig Inn was built by Lord Penrhyn in 1800 and soon became the most fashionable hotel in Snowdonia, offering spectacular views over the twin lakes of Llynnau Mymbyr towards the Snowdon hoseshore. Due to its popularity, the Shrewsbury to Holyhead mail coach (named the Ancient Briton) was re-routed away from the North Wales coastal route to run via Capel Curig in 1808. Many famous Victorians, including Queen Victoria, were to stay here to take in the celebrated scenery of the locality. In 1870 it changed its name to the Royal Hotel, and is today the main mountaineering centre for North Wales. (Courtesy of Plas y Brenin, the National Mountain Centre)

evening of Boxing day 1835, believed that: '… all except MD [Monkwearmouth Dock] resulting from the Clifton Bridge – which I fought hard for and gained only by persevering struggles and some manoeuvres (all fair and honest however)…'[48]

Brunel had also gained an important insight into the potential of the South Wales iron industry through his interest in ironwork generally and suspension chains in particular and, of course, rails for the GWR. He had made many contacts in the course of this work and one outcome was to be his first commission in South Wales.

Chapter 5 Notes

1 Rolt, LTC, (1959), p. *Isambard Kingdom Brunel,* Longmans, Green & Co., London,
2 Noble, Celia Brunel, (1938), p. 105, *The Brunels Father and Son*, Cobden-Sanderson, London.
3 Jones, Stephen K., (1981) p.37, 'A Link with the Past', in *Glamorgan Historian*, Vol.12, Denning, Roy, Ed, Stuart Williams Publishing, Barry.
4 Huchon, Rene, (1968) p.473, *George Crabbe and his Times 1754 – 1832*, Rene, Frank Cass & Co. Ltd, London, Diary entry for 24 October 1831.
5 Brunel, Isambard, (1870, reprinted 1971), p.46, *The Life of Isambard Kingdom Brunel,* Longmans, Green & Co., London, 1870, reprinted by David & Charles, Newton Abbot, 1971.
6 Buchanan, R. Angus, (2002), p.43, *Brunel, The Life and Times of Isambard Kingdom Brunel,* Hambledon and London, London, 2002. I am indebted to Prof. Buchanan for pointing that a cryptic entry in his personal diary that suggested he visited Redcliff in June 1828 is more likely 'Redriffe'.
7 Rolt, L.T.C., (1959), p.46.
8 Jones, Stephen K., (1981), p.37.
9 Body, Geoffrey, (1976), p.16, *Clifton Suspension Bridge An Illustrated History*, Moonraker Press, Bradford-on-Avon.
10 Todd, A.C., (1967), p.208, *Beyond the Blaze: A Biography of Davies Gilbert,* D. Bradford Barton, Truro. Diary entry for 23 April 1791, his paper on Suspension Bridges was read to the Royal Society in March 1826.
11 'On some Properties of the Catenarian Curve reference to Bridges of Suspension,' in a letter to the editor from Davies Gilbert Esq. F.R.S. and M.P. London Quarterly Journal of Science, 10, pp.230-35, 1820.
12 Todd, A.C., (1967), pp.99-100. Diary 30 January 1826 and note of 1839 quoted.
13 Buchanan, R. Angus, (2002), p.31, entry from the Thames Tunnel Journal 20-21 February 1829.
14 Rolt, LTC, (1959), p.56-57.
15 Noble, Celia Brunel, (1938), p.109.
16 Horsley, John Callcott, (1903), p.177, *Recollections of a Royal Academician.*
17 Gotch, Rosamund Brunel, ed, (1934), p.168-69, *Mendelssohn and his Friends in Kensington, Letters from Fanny and Sophy Horsley written 1833-35,* Oxford University Press, London.
18 Brunel, Isambard, (1870, reprinted 1971), p.56.
19 Brunel, Isambard, (1870, reprinted 1971), p.56.
20 Davies, John. (1981), p.223, *Cardiff and the Marquesses of Bute.* Cardiff: University of Wales Press.
21 Hilling, John, (1973) p.47, *Cardiff and the Valleys*, Lund Humphries, London. Edgar Jones in *A History of GKN* (Vol.1) names the architect as M'Culloch.
22 Webley, Derrick Pritchard, (1997), p.85, *East to the Winds, The Life and Work of Penry Williams (1802-1885),* The National Library of Wales, Aberystwyth. The architect was John Macculloch (1773-1835). See Kevin Littlewood's article in *The National Library of Wales Journal*, Vol.xxxi, Number 1, Summer 1999, *Rhymney's Egyptian Revival*, pp.11-39.
23 Greenacre, Francis and Stoddard Sheena, (1986) p.52, *The Bristol Landscape the watercolours of Samuel Jackson 1794-1869*, City of Bristol Museum and Art Gallery, Bristol.
24 *The Cambrian*, 25 June 1831, reporting on the event of 21 June 1831 in the *Bristol Mercury*.

25 Body, Geoffrey, (1976), p.33.
26 Huchon, Rene, (1968) p.474.
27 Rolt, L.T.C., (1959), p.58. An initial search of the Horsley family papers held by the Bodleian Library, University of Oxford, has also proved negative.
28 Humbert, Jean-Marcel & Price, Clifford, Eds, (2003), *Imhotep Today: Egyptianizing Architecture*, pp.58-59. UCL Press, London.
29 Buchanan, R. Angus, (2002), pp.53-54.
30 A long term exhibit at the Exeter Maritime Museum, the Bridgewater dredger has moved to the Bristol Maritime Museum.
31 GRO, D/D G Letterbook 1836(2) f.121, 26 October 1836.
32 Lady Charlotte Guest's Journal, 5 November 1836, p.399, quoted by Jones, Edgar, (1987) p.75, in *A History of GKN (Vol.1)*, Macmillan Press, Basingstoke.
33 Pugsley, Sir Alfred, ed, (1976), *The Works of Isambard Kingdom Brunel*, chapter on Railways by O.S. Nock, p.72, University of Bristol and ICE, (Brunel sketch book; SB, Miscellaneous).
34 Brunel archives, Brunel University, Letter Book 2, p.216, 11 January 1837, G.T. Clark to Dowlais Iron Company.
35 Sopwith, Robert, (1994) p.174, *Thomas Sopwith, Surveyor – An Exercise in Self-Help*, The Pentland Press, Bishop Auckland.
36 Glamorgan, Monmouth and Brecon Gazette and Merthyr Guardian, 20 July 1839.
37 GRO, D/D G LH 293, 14 August 1839.
38 GRO, D/D G LH 304, 6 September 1839 and D/D G LH 310, 16 September 1839.
39 GRO, D/D G LH 312, R.P. Davies to J.J. Guest, 21 September 1839 and D/D G LH 239, 10 June 1839.
40 Pugsley, Sir Alfred, ed, (1976), p.59.
41 GRO, D/D G Letterbook 1839 f. 548 15 August 1839 and 1840(2) f.233 14 November 1840.
42 GRO, D/D G LH 393 23 March 1840 and D/D G LH 439, 26 June 1840.
43 GRO, D/D G LH 180 Captain Richard Jenkyns to Messrs Guest, 11 September 1843.
44 Porter Goff, R.F.D., *Brunel and the design of the Clifton Suspension Bridge*, Proceedings of the Institution of Civil Engineers, Part 1, 56, pp.319-20.
45 Barlow, W.H., (1867) *Description of the Clifton Suspension Bridge*, Minutes of the Proceedings of the Institution of Civil Engineers, 26, pp.243-57. This paper has been reproduced in the first issue of a new journal; 'Bridge Engineering', part of the Proceedings of the Institution of Civil Engineers, edited by Professor Ben I.G. Barr, Vol.156, Issue BE1, March 2003.
46 Information supplied to the author by D. Mitchell-Baker of Howard Humphreys & Partners, Consulting Engineers, Leatherhead, March 1983.
47 Sir Allan Quartermaine, (1959), *I.K. Brunel – The Man and his Works, British Railways* (Western Society) London lecture and Debating Society, Session 1958-59, No.456.
48 Rolt, L.T.C., (1959), p.84, 26 December 1835.

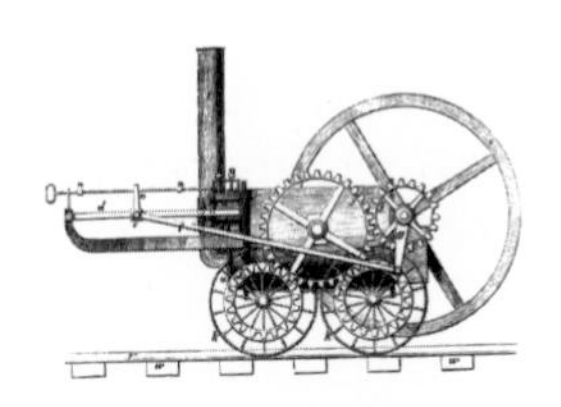

6

Railway Resurgence

'I Forgot Whether it was called a Tramroad or Railroad Company'[1]

Railway proposals in South Wales following Trevithick's run were not confined solely to tramroads; the example of the Stephenson-inspired line at White Rock has already been mentioned. Around the time of the Stockton & Darlington Railway, George Stephenson attempted to capitalise on the interest and potential demand for railways in several ways. The most important move was to establish a locomotive manufactory: Robert Stephenson & Co., but there was also a railway construction counterpart called George Stephenson & Co. One of the proposals put forward by George Stephenson for his company was for a 'London and South Wales Road', in 1824. Hugh Steel, the son of John Steel, who had been involved with Trevithick in 1804, was to assist with the survey. Hugh Steel was also to assist with surveys of the Liverpool & Manchester Railway but tragic circumstances were to affect both father and son.[2] The London and South Wales Road was one of the many lines projected by George Stephenson & Co. to try and monopolise railway construction, and George Stephenson even attempted to persuade Nicholas Wood not to publish his 'Practical Treatise on Rail-Roads' so that he could maintain the position he had as a railway engineer.[3] His London and South Wales proposal was to come to nothing although a revival, in terms of the route, was to coincide with the proposed railway between Merthyr and Cardiff.

Brunel had become known to the Merthyr ironmasters through his connection with iron work for his various projects, such as the Clifton Suspension Bridge. One of the popular stories regarding Brunel's introduction to the subject of a railway between Merthyr and Cardiff concerns Anthony Hill of the Plymouth ironworks: 'Anthony Hill asked Isambard Kingdom Brunel, who was a personal friend of Hill's, to estimate the cost of a railway from Merthyr down to Cardiff.' This was Barrie's view, written in the 1969 edition of his book on the Taff Vale Railway, which is different to the first edition of his work (1939) in which he highlights Guest's role.[4] Certainly, Hill and Brunel were to become friends who corresponded with each other throughout their respective careers, and may have first met in connection with Clifton.

Brunel was actually given the go-ahead for a survey of a railway between Merthyr and Cardiff by Alderman William Thompson (*d.*1854) on 7 October 1834.[5] That

Thompson was involved in this way revives the notion of the old triumvirate of Penydarren, Dowlais and Plymouth ironmasters in the development of a new transport initiative, as they had done so in the promotion of the Merthyr Tramroad some thirty-five years previously. Brunel himself had identified the source of this commission when later wrote in his diary: '... Merthyr & Cardiff Railway - This too I owe to the G.W.R...'[6] What Brunel meant by this was that the TVR commission had come his way because of the profile he had gained (and contacts made) as the engineer of the proposed GWR. Local newspapers carried reports on the railways development, and advertisements appeared following the issue of the first prospectus of the GWR outlining the share capital available. One such advertisement appeared in the *Glamorgan, Monmouth and Brecon Gazette and Merthyr Guardian* and *The Cambrian* newspapers on 12 October 1833. In the following weeks both newspapers were to report on meetings for a proposed Cambrian, Gloucester and London Railway, which sought to join the projected railway from Bristol to London.[7]

The Cambrian was quicker off the mark, reporting this meeting in its edition of 19 October 1833, after the meeting had taken place at the 'Castle Inn', Merthyr, on 11 October 1833. This meeting was dominated by ironmasters. Indeed, the proposed committee for the railway following the adoption of the resolution (below) included every Welsh ironmaster of note, including: J.J. Guest, Anthony Hill, W. Thompson, Samuel Homfray and William Crawshay.[8] Guest took the chair and briefly opened the business of the meeting, making the following statement: '... a railroad on the line now proposed, or on one nearly similar, must be at no distant time adopted, and the question was whether this was the time it may be successfully attempted.'[9] At the meeting the question was raised whether this proposal would clash with the projected Bristol railway. Guest believed it was in the interests of the GWR to form a junction with the line proposed. The engineer behind the Welsh proposal was William Wooddeson, who was described as a surveyor and architect, but what experience he had in these matters was not given. It was resolved, 'That the formation of a Railway from South Wales through Gloucester, to join the projected railway from Bristol to London, with a branch from Gloucester to Birmingham, would be of great national importance...'[10]

A similar meeting was held at the Town Hall in Swansea on 17 October. There, one of the industrialists present succinctly commented on the main problem facing the proposal: '... raising the "sinews of war", money, money, money was the only difficulty.'[11]

Raising the money required may not have been the only difficulty but it was certainly the major obstacle facing the proposed Cambrian, Gloucester & London Railway. The following week's edition of *The Cambrian* (26 October) carried a letter on this subject, stressing that the fear of obtaining funds should not stop the westward extension of the line at Swansea, '... to save money'. It was signed 'W' and was possibly Wooddeson's reaction to the obstacle. Editorial comment in the *Glamorgan, Monmouth and Brecon Gazette and Merthyr Guardian* was, like *The Cambrian*'s, positive: 'We congratulate our fellow citizens...' [12] Meetings continued and the title of the proposed line changed to reflect further opportunities, such as on 5 November 1833 when it was being styled the Cambrian, Gloucester, Birmingham & London Railway, and Wooddeson was described as a civil engineer.[13] However, on 16 November *The Cambrian* reports on a meeting held in Gloucester in which '... very serious objections appear to have arisen to the proposed line of road, another General Meeting is to be immediately holden, to take such objections into consideration.' By the end of the year, Wooddeson appears to have disappeared from the scene, defeated – it would appear – by the 'sinews of war'. *The Cambrian* summed up the situation the following year (5 April 1834): 'When Mr. Wooddeson, in October last adduced his plan for a Railway from London via Gloucester and Swansea to Milford Haven, a general impression prevailed among our most influential men of business, that such a scheme could only be effected by Government...'

This did not mean that these 'grand' proposals for South Wales were entirely dead; in the same article *The Cambrian* referred back to the previous week when a proposal for a Llanelly to Cardiff line, as part of a grand section of the GWR, was mooted, and two weeks later carried a notice for the 'Grand Cambrian and Western Railway'. Henry Habberley Price was involved with this proposal and had previously been involved with William Brunton in putting forward the first practical proposal for a line between London and Bristol in 1832 (see Chapter 4).[14] Price was to later support Brunel's route during the reading of the 1835 GWR Bill, and it is clear that Brunel and other supporters were lobbying all influential parties to protect the GWR's position, and to ensure that any 'grand sections' forming a junction with the GWR were under its influence. A further proposal, the 'Gloucester and South Wales', was put forward as part of the GWR grand plan, although this was to ultimately fail on financial grounds.

The GWR itself did not have an easy ride. Following Parliament's rejection of the first Bill in July 1834, a new Bill was drafted and was heavily promoted during the autumn and winter of 1834/35. MacDermot refers to the promoters as being: '... busy stirring up support, financial and moral, all over the West of England and in South Wales'.[15] The importance of the railway as being part of a future route to South Wales was therefore heavily promoted and a projected line from Merthyr to Cardiff was shown on the map accompanying the second prospectus.[16]

Like Guest, who was an MP with industrial interests in Merthyr, Alderman William Thompson was keen to promote improved communications. In 1819 he had become joint proprietor of the Penydarren ironworks, taking control following the death of William Forman in 1829. As well as twice being Lord Mayor of London, he was MP for the City of London, and had married one of Samuel Homfray's daughters, so some thirty years after a Penydarren ironmaster had commissioned Trevithick to build the first railway locomotive, another Penydarren ironmaster was to commission Brunel to build a railway. Thompson was also no stranger to railway matters; between the years 1826-30, the Penydarren ironworks had sold iron to the value of £1,252 to the locomotive manufactory of Robert Stephenson & Co. It is possible that engineering products may have been received in part payment for these transactions as a Stephenson 10hp stationary engine was purchased in 1827, and a 0-6-0 locomotive delivered in 1829.[17] This was the locomotive that was later used on the Merthyr Tramroad in a rebuilt form as the 0-4-0 engine *Eclipse*. In 1830 the Penydarren ironworks supplied rails to the Liverpool & Manchester Railway. With such strong Stephenson connections it is surprising that they were not invited to build a line down in Wales but, as Brunel states: '... Merthyr & Cardiff Railway – This too I owe to the G.W.R...'[18]

It is possible that there were other, and earlier, connections between the ironmasters of Merthyr and Brunel, through Brunel's connection with the London scientific circle and individuals such as Michael Faraday and Charles Babbage[19] or the wider political community. Brunel's brother-in-law was Benjamin (later Sir) Hawes (1797-1862), who had become MP for Lambeth in the post-Reform election of 1832. Guest had first entered Parliament in 1826 (for Honiton in Devon), with the passing of the Reform Act, Guest stood for Merthyr and was returned, a position he was to occupy until his death in 1852. Thompson was MP for the City of London and he and Guest moved in the same circles. On 8 May 1834, Lady Charlotte Guest records her experiences of a London dinner party hosted by Thompson:

> 'Conceive the horror of seeing a fat woman sit opposite to one in a yellow gown, and an amber cap with red flowers, and the still greater horror of that fat lady claiming to be an acquaintance. She proved to be Mrs Hudson...The Browns [Major Gore Brown] and the Hudsons were almost the only two families in London that I have taken pains to avoid...' [20]

'Castle Inn', Merthyr Tydfil. (Courtesy of Alan George)

This was the wife of George Hudson, the 'Railway King' whose railway empire was to spectacularly collapse in 1847, but in the meantime was the sort of man that iron-masters such as Thompson and Guest relied upon for their business. Lady Charlotte's background was considerably more refined than Mrs Hudson's, but her reaction comes across as being amazingly snobbish considering her relationship with Guest. Despite the fact that Guest was one of the richest industrialists in the country, and was to leave £500,000 and the vast country estate of Canford Manor in Dorset in his will, there was always the charge levelled against her that she had married a 'man in trade'. London society had looked with horror upon the second marriage of a forty-eight-year-old ironmaster with the twenty-one-year-old eldest child of the ninth Earl of Lindsey. To some, Mrs Hudson could be accused of being one of the *nouveau riche* with bad taste, but not of betraying her class. Charlotte was determined, however, not to remain in her husband's shadow; she would not only make her way in society, but would fully embrace her role as the wife of the largest manufacturer in the world and, indeed, go beyond that and actually run the vast industrial empire that was Dowlais. In the middle of all this and raising ten children, she found time to learn medieval Welsh and translate the medieval Celtic stories known as the *Red Book of Hergest* and the *Hanes Taliesin*, which she was to publish as the *Mabinogion*, the publication of which would even have an impact on the affairs of the Taff Vale Railway.

One impression of Charlotte has been left by Lady Holland, at that time the grand lady of society, who had once famously agreed to make a journey by train on the condition that Brunel would accompany her and hold her hand;

> ''I have got acquainted with a very remarkably clever, distinguished woman, reckoned by many extremely handsome, Ly C. Guest, nobly born, married to an immensely rich man, who wanted what the Spaniards call Sangre Azul, and gave her wealth which she wanted. They seem perfectly happy; his riches are in Wales.'[21]

Lady Charlotte Guest records a meeting of Brunel with Guest, in connection with the Merthyr to Cardiff Railway, in 1834:

> '... October 12... Mr. Brunel of the Thames Tunnel, accompanied by Mr. Frere, came here in the evening. They are to make a survey of a railroad from Merthyr to Cardiff, and Merthyr [her pet name for her husband] got up soon after six this morning, in order that he might have a very early meeting with them on the subject...'[22]

The Frere mentioned here is George Frere, who can claim a Welsh connection through his father's partnership in the Clydach ironworks in Breconshire. Although he was one of Brunel's chief assistants at this time and became the resident engineer for the GWR's Bristol Division, he was to part company with Brunel, on less than amicable terms, in August 1841.[23]

The time appears to have been right for railway investment between Merthyr and Cardiff. In 1834, the Penydarren ironworks sent some 12,752 tons of iron down the canal, slightly more than Plymouth who, in the same year, sent 12,073 tons. Both sent considerably less than Cyfarthfa (34,952 tons) and Dowlais (33,072 tons).[24] Events would show, however, that interests other than that represented by ironmasters needed to unite behind such a proposal to make it a reality. Brunel's £190,649 estimate for the construction of the line was accepted, and he undertook the survey for the line. In his office diary, Brunel records subsequent visits to South Wales; on 22 October 1834 there is a margin note M & C [for 'Merthyr & Cardiff']; and, 'engaged on survey – met Mr A Hill & Mr J Guest at Cardiff – 1 day. Left Frere at Cardiff.'[25] On the same day he also allocates a half day to GWR business 'correspondence', and returns to M&C business the following day (Thursday): 'Engaged in survey from 5 a.m. – at Cardiff – met Cubitt – returned late – 1 day Travelling all night – ½ day.' This was William Cubitt, the dock engineer who was advising the Marquess of Bute on the construction of his dock from 1833. The following day (Friday) he was again engaged on M&C business, allocating half a day for travelling and a note of expenses for 'Frere's acc[oun]t'. The weekend was spent on GWR business, but there is a return to M&C business on Monday 27 October 1834.

Plans from this survey were deposited and a notice of the Bill appeared in the local press: 'Merthyr Tydfil and Cardiff Railway…application is intended to be made to Parliament in the next session for leave to bring in a Bill to make and maintain a Railway with proper works and conveniences connected therewith…' However, no action was actually taken until the following year.[26] The shipping destination is given as the proposed ship canal (Bute Dock) and there is no reference to an Ely branch in this notice.

John Callcott Horsley, 1817-1903. Horsley was a student member of the Royal Academy of Arts when he travelled with Brunel. He was elected A.R.A. in 1855, R.A. in 1864 and served as treasurer from 1882-1897. He is credited with the design of the first commercial Christmas card in 1840. His prudish attitude to nude studies earned him the nickname of 'Clothes-Horsley'. (SKJ collection)

On some of these and subsequent surveying visits to Wales, Brunel was accompanied by John Callcott Horsley, who later recalled his experiences of travelling with Brunel:

> 'I often accompanied I.K.B. on his journeys to the South Wales iron districts, where he was actively engaged in making the railway from Cardiff to Merthyr Tydfil. The vast work of the GWR with its network of branches loomed high upon the horizon of his manifold labours, and the power for work that he showed at this time was almost incredible. I was with him in many of these Welsh journeys, which he made in a large britzska, known to the post-boys as the "Flying Hearse" from its unusual dimensions, and the speed in which it was whirled along night and day by four horses. He was a perfect travelling companion, save for one most trying habit with which nature had endowed him beyond that of any other human being with whom I ever foregathered in sleeping hours. His potency in SNORING was "prodigious" as Dominie Sampson would have said, and in our night-long travellings I had to shake and pommel him to that extent that I constantly anticipated his turning and rending me in his sleep as an actual assailant.' [27]

The reference to his 'power of work' is also used by G.T. Clark: '... I never met his equal for sustained power of work...'[28] Clark had supplied some reminiscences for Isambard Brunel to use in his biography, and also refers to his mode of travelling:

> 'When he travelled he usually started about four or five in the morning, so as to reach his ground by daylight. His travelling carriage, in which he often slept, was built from his own design, and was a marvel of skill and comfort.' [29]

The night-long travellings are certainly borne out by the entries in his office diary. His 'power for work' is also evident in Horsley's remarks, particularly with regard to the deadline Brunel faced in the completion of the survey:

> 'On one occasion he was completing the survey for the Taff Vale Railway, of which the plans were required to be sent to the headquarters of the Railway Company by the 30th of November. For a fortnight Brunel travelled about day and night without once going to bed, getting all the sleep he had in the Flying Hearse. I was quartered with friends a few miles off, and used to ride in daily on a Welsh pony to get my orders. On the last day of the fortnight, when I rode in early, I heard that the plans had been delivered at 3 a.m., and that Mr. Brunel had then gone to bed. After a time I entered the room with great caution, and found him in a deep sleep. He was in the habit of smoking a cigar the last thing at night, and one was lying across his chin, one end in his mouth, the other showing signs that it had been lighted, though not a solitary puff had been smoked, he having sunk into the sleep of the just the moment his head touched the pillow. He woke naturally after exactly twenty-four hours of profound repose, and we started for London.'[30]

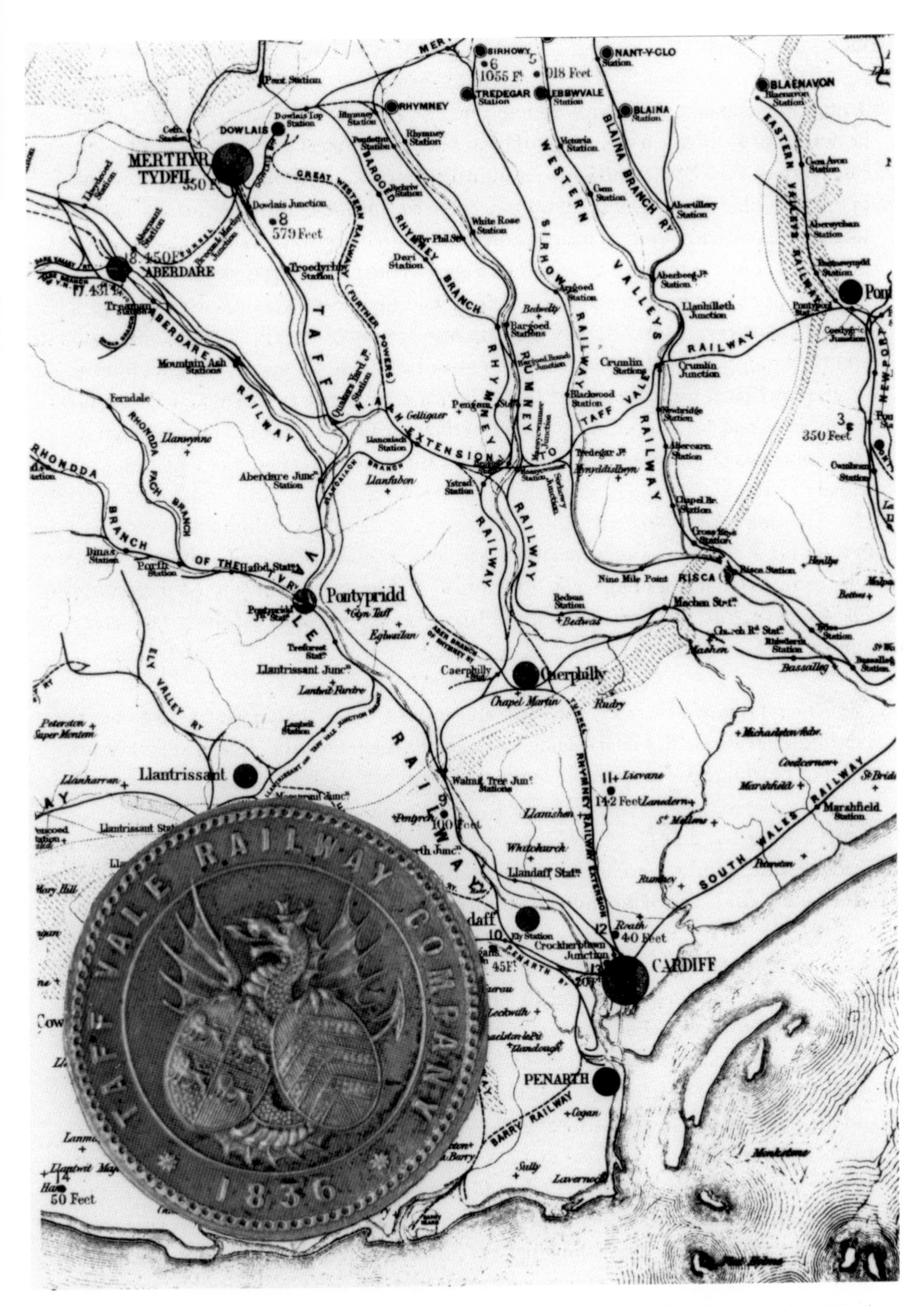

A map showing the TVR main line between Cardiff and Merthyr Tydfil and branches, with the original TVR seal. (TVR seal courtesy of GWR Museum, Swindon)

It was not long before the proposal was revived. Brunel wrote to Guest on 9 October 1835, following a revival of interest in the MT & CR: '… This is a most busy time of year, and I had dismissed the idea of the Merthyr & Cardiff.'[31] He went on to enquire: 'If you are thinking of coming to Parliament this year, I should certainly wish to have some conversation with you about the terminus at Cardiff, and also about the connecting branches at Merthyr…' The revival was linked to developments following the breakdown of negotiations regarding the Glamorganshire Canal. In June of that year, R.J. Hill, T.R. Guest, Thomas Charles and Walter Coffin left the canal committee and, late in the summer of 1835, the major users of the canal held a meeting, at which there was much criticism concerning the shortcomings of the canal.[32] Hadfield refers to a second meeting in which William Crawshay II, acting on behalf of the canal company, proposed a programme of improvements which cost in the region of £30,000 to £40,000. It was not enough for some; to quote Hadfield: '… what no one had the courage to say face to face with the Crawshays, some were determined to do – they went on with the railway scheme.' [33] That determined group, which included not only ironmasters (it was to be chaired by Sir John Guest) but also colliery owners such as Walter Coffin, Robert Beaumont and Thomas Powell (1784-1864) were to meet on 12 October 1835 at the 'Castle Inn', Merthyr. Here the Taff Vale Railway was to be formally inaugurated (the new proposal was to be given a new name rather than the Merthyr Tydfil & Cardiff Railway), a second attempt achieved through the union of those two interest groups, ironmasters and coal freighters, that used the Glamorganshire Canal. A report of the meeting noted that:

> 'It was well known that some negotiations had been lately attempted on the part of the freighters with the GC Co., Those negotiations had been broken off in consequence of the Canal Co having withdrawn the proposals they had made at a previous meeting to carry on certain terms.' [34]

Thomas Guest addressed the meeting, pointing out the benefits that such a railway would bring, and referred to the recent stoppage of the Glamorganshire Canal for ten days. He used the Stockton & Darlington Railway as an illustration, stating that it,

> '… was a far more expensive line, in every respect, than the one now proposed, according to the representation of Mr Brunell, [sic] the respectable engineer of the GWR, and they had a right, therefore, to calculate on a return of profits fully equal to the most sanguine expectations of the promoters of the Rail-road.' [35]

The reference made by Thomas Guest to '… return of profits…' indicates the wider economic forces that were beginning to drive the interest in such an undertaking. That week's edition of *The Cambrian* (17 October 1835) highlighted the growing demand for railway shares, pointing to the success of the Manchester & Liverpool Railway as the principal cause. While not in the same league, shares in the TVR were to prove as popular. A week later, the same newspaper reported that, 'The Taff Vale

Railway Share list was closed on Wednesday last – the whole having been taken in so short a time after the prospectus was issued.'[36] A week after that it commented on the fact that the whole of the Gloucestershire & Birmingham Railway shares were appropriated and that the TVR shares were already at a premium of £5 '... and scarce at that price.' [37]

The TVR was caught up in what was to become known as the 'little' railway mania of 1834-37. Newspapers such as *The Cambrian* and financial institutions took stock of the situation; '... there is at present a great madness abroad in regard to railroads...' was one comment from a partner in Baring Brothers, the merchant bank that survived this mania, the later 'large' railway mania of 1845-48, and indeed all the market crashes until it was brought down by the 'rogue trader' in the 1990s.[38] The Manchester & Liverpool Railway was returning a dividend of 10 per cent, a performance also matched by the Stockton & Darlington Railway. Such returns from railway shares represented an opportunity for profit unmatched anywhere else; there was no comption with the relatively low yields available in the bond market. It was, however, the initial investor that reaped the 10 per cent dividend, not the buyer in the secondary market if the shares went to a premium. Speculators were creating a demand for railway shares and new issues, in particular those relating to major trunk routes. The GWR was to be very popular; in September 1836 5,000 shares attracted 8,000 applications – these were £100 shares on which £5 had been paid, and were now changing hands at £8, and even £11 to £12.[39] A steady fall in interest rates fuelled this demand for high yield investments that the railways were delivering. However, it was short lived: in May 1835 the GRS Railway Share Index stood at 60. It peaked at around 130 a year later, and fell back to under 80 by April 1837.[40] Even after the bubble had burst, railway construction continued for some time and, although a recession followed, the railway mania was not felt to have been the principal cause; the 'large' railway mania was to have far more serious consequences.

In the edition of the *Glamorgan, Monmouth and Brecon Gazette and Merthyr Guardian* that reported the meeting at the 'Castle Inn' (17 October 1835), a notice carried three resolutions, the first resolving that the present means do not afford the requisite facilities, the second, 'That it is expedient to establish a communication by means of a Railway, which shall combine the advantages of the latest improvements in the mode of transport.'[41]

This was to be achieved through the third resolution, namely the formation of a company to be called the 'Taff Vale Railway'. The provisional committee formed following the meeting also resolved to connect various collieries to the railway, such as those belonging to Thomas Powell and Sir Christopher Smith in the Llancaiach district, and Robert Beaumont's colliery in the Bargoed area, that is '... after a further view of the locus by Mr Brunel' and a branch into the Rhondda to 'terminate at or near the end of Mr Coffin's tramroad.'[42] An outlet to the sea at the mouth of the river Ely and access to the proposed ship canal (Bute Dock) formed other resolutions, all of which were to result in significant changes to the 1834 proposal. *The Cambrian* carried the notice of Bill for the Taff Vale or Merthyr Tydfil & Cardiff Railway in its edition of 21 November 1835, listing those changes, and the intention to construct its own shipping facilities and additional branch railways,

> '... and also another Branch Railway, with proper wharfs, staiths, works and conveniences connected therewith, leading from and out of the said first-mentioned Railway, to or near Cogan Pill, on the west bank of the River Ely, in the Parish of Landough, and County aforesaid...' [43]

The branch railways included, in addition to alterations to the Merthyr branches, the Llancaiach branch and the Dinas branch into the Rhondda, where it was to connect with the Dinas Tramroad at Aerw Isaf. Obviously the last two branches served those colliery proprietors supporting the TVR proposal: Thomas Powell (Llancaiach) and Walter Coffin (Dinas). In the same edition of *The Cambrian* it was reported that a company had been projected to establish a steamship connection between Bristol and America. This would appear to be the Great Western Steamship Co., which was formed in Bristol as the result of Brunel's interest in a steamship extension of his GWR to America. On the previous day (20 November 1835), Brunel had written to Henry Scale of the Dowlais Iron Co., seeking payment of his account for 'expenses of last year' amounting to £500, presumably his survey costs.[44] There were also the expenses of an assistant, Mr Johnson, for whom Brunel sought the payment of £200 from the Merthyr Tydfil & Cardiff Railway Co. Edward John Hutchins, in turn, sought payment from Thomas Revel Guest at Dowlais on 25 November 1835: 'My Dear Uncle, Mr. Brunell [sic] has written to request me to pay Mr Johnson for £200...'[45] That Brunel was heavily engaged on the GWR and now had a steamship to design were possible reasons why he had overlooked his account in respect of this survey, a fact he admits was his fault to Henry Scale. It would appear to sum up his regard for the Merthyr Tydfil & Cardiff Railway (what Mr Johnson thought of being kept waiting so long for his payment is not recorded), but what is even more interesting is Brunel's postscript: 'I forgot whether it was called a tramroad or railroad company. I have dignified it with the latter appellation.'[46]

The fact that Brunel was unsure whether Merthyr Tydfil & Cardiff Railway was to be a tramroad or railroad company is most telling. Some ten months later there is a letter from Brunel to George Bush, in which he talks about having postponed his visit to Merthyr until Saturday (Bush not having time to see him on the Friday), and that he will come across by the Cardiff boat from Bristol, picking up Bush at the 'Cardiff Arms' in Cardiff to proceed to the 'Castle Inn' in Merthyr,

> '... where I hope to find the plan and sections of the proposed branches north of Merthyr ready & before I leave which I trust will be on Monday Even I hope to have the survey of the railway finished that I may examine it. [47]

Brunel must have been in no doubt by this time about what kind of railway the TVR would be, but closer examination of the last line of the original document shows that the word 'railway' appears not to have been written in full, with 'rail' looking as if it was added at a later date. In the long titles of Acts of Parliament relating to Wales, the Taff Vale Railway's Act of 1836 (6 Wm. IV, cap lxxxii), 'An Act for making a Railway from Merthyr Tydfil to Cardiff to be called "The Taff Vale Railway" with branches', does not mention tramroads whereas the Llanelly Railway and Dock Co. does.[48] This is borne out by the full description of the TVR Bill which talks about railways, branches or branch railways and only mentions tramroads in the context of, '… a Branch to communicate with the Tramroad leading to the collieries called Dinas.' The Llanelly's Act was granted the year before (1835) and, as a Welsh railway company, is immediately prior to the TVR's. It mentions that the company has the power to make: '… certain additional Railways or Tram-roads…' From the TVR onwards, there are no further references to tramroads in the long titles of railway companies, even in the case of subsequent Acts for existing tramroad operating companies, such as the 'Duffryn Llynvi and Porth Cawl Railway'. In view of the time that Guest and others must have spent with Brunel concerning the railway, the comment about what kind of company it should be is intriguing, and may suggest that Brunel was unsure if the company would also wish to construct tramroads e.g. for the minor quarry branches.

With the formation of a company, at which Guest was to take the chair for the first meeting, a full survey of the route to be undertaken was authorised. An early entry for 1836 in his Letter Book gives Brunel's revised estimate for the construction of the line, now considerably more than his first estimate, to accommodate improved gradients, mineral branches and shipping staithes. This estimate came to £286,031.18.00, breaking down as follows:

Estimate[49]	£	S	D
Earthwork	83,333	.0	.0
Tunnelling and Masonry	38,696	.0	.0
Forming Road	86,000	.0	.0
Land and Depots and compensation	40,000	.0	.0
Staiths and shipping places	8,000	.0	.0
Machinery	4,000	.0	.0
	260,029	.0	.0
10 per cent contingences	26,002	.18	.0
	286,031	.18	.0

Taking stock of the situation, late on the evening of Boxing day 1835, just two years after his appointment to the GWR, Brunel was to write in his diary:

> 'The Railway now is in progress. I am their Engineer to the finest work in England - a handsome salary - £2,000 a year - on excellent terms with my Directors and all going smoothly, but what a fight we have had - and how near defeat - and what a ruinous defeat it would have been. It is like looking back upon a fearful pass - but we have succeeded. And it is not this alone but everything I have been engaged in has been successful.'[50]

In listing his 'stock of irons', Brunel records the Taff Vale Railway but it is clearly not in the same league: 'Merthyr & Cardiff Railway – This too I owe to the G.W.R. I care not however about it'. Returning to the diary entry, we find that Brunel was to list these 'jobs', as he describes them, and the capital investment they represent:

Capital	70,000	Clifton bridge
	20,000	Bristol Docks – to come – Portishead Pier
	2,500,000	G.W. Railway – to come – Oxford Branch
	750,000	Cheltm Railway
	1,250,000	Bristol & Exeter do.do.- perhaps Plymouth etc
	250,000	Merthyr & Cardiff do. Gloster & S. Wales
	150,000	Newbury Branch
	50,000	Sunderland Docks
	100,000	Thames Suspension Bridge
	450,000	Bristol & Gloster Railway
	5,320,000	

Brunel may not have regarded it as an important commission but the Taff Vale Railway was to be an exceptional case in other ways. The difference in rail gauge meant that there would be little common ground between the GWR and TVR, and future connections between the TVR and the broad gauge would prove to be less than straightforward. Unlike the majority of Brunel's railway projects which become absorbed into the GWR at an early stage, the TVR would not become part of the GWR, and then as a constituent company, until grouping in 1923.

With progress being made on the construction of the line the inevitable inflated demands for compensation and 'incentives' to remove objections began to come in. One of these related to existing commercial and property interests in the area. In January 1836 Brunel received a letter from Mr Rickards about the compensation which he expected for his property and a tramroad near Newbridge (Pontypridd). Brunel told the directors:

> 'It's a mere question of money and I think he had better be bought if he can be had cheap.... We go through houses at Newbridge for which of course we must pay. The Tramroad is a clear case of competition and cannot I suppose be remedied.'[51]

He wrote politely to Rickards, suggesting that he would get more accurate advice from the railway company, care of either E.Hutchins Esq. of Dowlais, or their Solicitors, Messrs Swaine, Stevens and Co. of London. One objection to the Bill was overcome with a £10,000 'accommodation' consisting of Exchequer Bills to be paid to the Glamorganshire Canal by Richard Blakemore of the Melingriffith works.

The following Bill was laid before Parliament in 1836:

6 WILL. IV.--SESS. 1836.

A

BILL

For making a Railway from *Merthyr Tydfil* to *Cardiff*, to be called "THE TAFF VALE RAILWAY," with Branches.

WHEREAS the making a Railway from *Merthyr Tydfil* to *Cardiff*, with Branches therefrom, to or near the several Iron Works of *Cyfathfa*, *Pen-y-darran*, *Dowlais* and *Plymouth*; the Limestone Quarries, called *Morlais* Castle and *Twynau Gwynion* and the Collieries called *Lancaiach*; and also a Branch to communicate with the Tramroad leading to the Collieries called *Dinas*; and another Branch leading to or near *Cogan Pill*, on the west bank of the River *Ely*, all in the County of *Glamorgan*, would be of great public convenience by opening an additional certain and expeditious means of conveyance to the Sea for the extensive mineral and other produce of the Places and Works above mentioned; also for the Carriage to the said Works from the Port of *Cardiff*, of iron ore and other mineral produce used in the fabrication of iron; and for the conveyance of Passengers and Goods to and from the said Towns of *Merthyr Tydfil* and *Cardiff*, and the several intermediate and adjacent Towns and Districts: Preamble:

And whereas the several Persons hereinafter named are willing, at their own expense, to carry into execution the before-mentioned Undertaking: BUT the same cannot be effected without the Authority of Parliament;

10. A May

1836 Taff Vale Railway Bill. (SKJ collection)

Despite canal opposition, the company was granted an Act of Parliament on 21 June 1836, and the actual Act included several alterations and amendments to the above Bill. The proposed branches to the ironworks and the limestone quarries were dropped and several clauses were added, including clause clviii, which limited the speed of trains to 12mph. This was repealed in 1840.

The first General Meeting of the proprietors following the Act of Incorporation was held at the 'Angel Inn', Cardiff, on 16 September 1836. Walter Coffin was to take the chair and a number of resolutions were passed. No.5 related to the appointment of twelve directors under the 94th clause of the Act of Incorporation. These were: J.J. Guest Esq. MP; Walter Coffin Esq.; T.R. Guest Esq.; Thomas Powell,Esq.; Thomas Carlisle Esq.; E.H. Lee Esq.; Henry Rudhall Esq.; C.E. Bernard Esq.; Christopher James Esq.; W.K. Wait Esq.; E. Waring Esq.; R.H. Webb Esq. The next resolution related to their remuneration, for which £500 was to be annually appropriated for the services of the directors, '... to be applied in such manner as they may consider equitable and most conducive to the interests of the Company.'[52] The meeting then turned its attention to shipping on the river Ely, to obtaining all requisite information on constructing docks, to procure plans and estimates ready for a special General Meeting and to apply to Parliament. Thanks were recorded to E.J. Hutchins for acting as Hon.Secretary to the company, and J.K. Brunel [sic] and W.O. Hunt. Under the financial statement, the engineer's account, including 'Professional Evidence in Parliament and cost of materials for survey', amounted to £1,902 13*s* 5*d*. The *Glamorgan, Monmouth and Brecon Gazette and Merthyr Guardian* reports that 'appropriate speeches' were made, including one by Brunel. The attached report expanded on the arrangement with Richard Blakemore and an agreement reached with the Dowlais, Penydarren and Plymouth companies to purchase the Merthyr Tramroad for £21,000. This would release the three companies from a Deed of Covenant, '... under which they were bound to pay tonnage upon the Old Tramroad, whether they did or did not use it.'[53] Despite the report saying that this arrangement would not be disadvantageous to the company as the sum would be offset by the cost of making a branch railway to the Penydarren and Plymouth limestone quarries, the powers were not exercised.

Brunel's reputation at this time appears to have extended beyond the boundaries of railway engineering and he was even consulted on the construction of a new turnpike road. On 5 December 1836, Brunel wrote to W. Meyrick: 'I have not heard from Mr. Powell about the proposed Neath Rd., is it still the wish of the trustees for me to do anything... I advised Mr. Powell to procure the assistance of some other professional man and probably he has done so...'[54] A more personal engagement was proposed by the Guests to the Brunels the following month, but Brunel was forced to decline: '... I beg you will express the very sincere thanks of Mrs. Brunel and myself for Lady C Guest's kind invitation but I regret my movements next week will be too rapid to allow of my taking Mrs. Brunel with me.'[55]

A second Bill was taken to Parliament the following year, '... to enable the Taff Vale Railway Company to alter the Line of the said Railway and the Act relating thereto and to make additional Branches.'[56] In April 1837 Brunel was called to give evidence on this Bill; he was examined by Mr Wood and Mr Hildyard on aspects of the Bill ranging from the Merthyr ironworks branches, raising additional capital, a deviation on the original line (improving a curve on the Ely branch line), and a branch to the Melingriffith works. Brunel was cross-examined at length on the nature and cost of working the branches and the inclined planes on them, arrangements made with Richard and John Hill in respect of the railway crossing land in the vicinity of the Plymouth ironworks, sites of stations, and if the Ely branch was a new 'communication with the sea'. Mr Hildyard cross-examined Brunel on the progress made under the 1836 Act:

'Your Act received the Sanction of Parliament on 21 June last year?
Yes.
By the Powers of your Act you could only purchase within two years; have you purchased any land?
I do not know; I should say not.
You have suffered one of the years to go by without contracting for land?
Not one of the years. As I said before, there is an immense deal to do before we can determine exactly the proportion to be taken, and that is a portion of the time contemplated to be included within the two years.
All this surveying and boring, and so forth – do you not think that had better precede an application to Parliament?
Another 10,000l might be spent, of course, before we went to Parliament, and very beneficially, but I do not know how these gentlemen would like us to be sinking shafts in the ground and cutting down all those trees that might be in our way.
Have you been cutting down trees now?
Yes.
You have done that without contracting with the owners?
It is a very necessary Power.
To what extent have you cut down trees?
Wherever they have obstructed our line of vision.
Have you put a spade into the ground?
Certainly; a good many spades.
You understand the phrase?
If you will, let me put it in my way. Contracts for large works are only just out. I have just reported on a tender, and whether it is actually let or not I do not know. The parties tendered to do it; I reported as to which was to be accepted, and that man is bound to do it. Whether his contract is signed, or whether it is being prepared to be carried into effect, I do not know.
There is a definitive contract yet entered into?
Yes.
What is the name of the contractor?
Jones, I think.
What is the extent of that contract?
It is a large bridge at Newbridge, halfway up the line.
When was the contract advertised?
A short time back. It has been all done in the usual course of things. That is the only contract for which a party's offer has been taken.
And you have commenced none of your embankments or cuttings, or works of that description?

No.
Nor contracted for them?
No.
Nor contracted with a single individual for the price of the land?
That I cannot speak to. We had agreements by the Act with some of them.
But those are conditional only if you enter upon the land?
Some of them are positive agreements.
Do you ever agree with individuals except with a proviso if you enter upon their lands?
I do not know whether we have in this case or not; there are so many agreements in all these Acts that I do not remember, but I can speak to many instances in which we have agreed to purchase, and where money has been deposited.
When do you contemplate that this main line will be opened?
The whole distance?
Yes?
Not under Three Years.
From this time?
Yes.
And how long will it take to make the extensions?
We shall do it so as to finish within the time.
The witness is directed to withdraw.'[57]

Did this performance come up to that which his friend, St.George Burke QC, alluded to when he made the following observation of Brunel under cross-examination?

> 'He was almost as much of a diplomatist as an engineer, and knew perfectly well how to handle a case in the witness-box so as to leave no loophole for his opponents to take advantage of. At the same time he was a perfectly honest witness, and while he avoided saying more than was necessary for the advancement of the case in which he was engaged, he would have scorned to say or imply anything by his evidence inconsistent with strict truth.'[58]

Walter Coffin also gave evidence to the House of Lords committee, confirming that there were no works in operation now, and that surveys had taken up a considerable time: 'I believe contracts have been advertised for, and I think one contract is taken...'

The bridge mentioned, the first contract awarded on the TVR, was the Newbridge Viaduct at Pontypridd, Pontypridd being known as Newbridge until the 1860s. The viaduct as an engineering work, however, is still officially listed in engineering records as the Newbridge Viaduct, and was Contract No.2, 'The construction of a Stone Bridge across the river Rhondda at Newbridge.'[59] That he was unsure of the contractor's name indicates the level of commitments Brunel was engaged in, and that he needed help in engineering the TVR. This was to be found in the shape of a competent resident engineer, George Bush (*c.*1810-41).

Little biographical information survives on George Bush, due to his early death in November 1841, a little over six months after the Taff Vale Railway was opened between Cardiff and Merthyr. He did not join the Institution of Civil Engineers during his short career, and no professional obituary has come down to us.[60] The foundation for his experience appears to have been gained in dock and harbour engineering, under one of the leading civil engineers of his day; Alexander Nimmo (1783-1832). As an engineer, Nimmo was responsible for surveys in Scotland and Ireland, and he directed the construction of some thirty piers or harbours upon the Irish coast, designing the Wellesley Bridge and docks at Limerick. Towards the end of his career, he was engaged in railway projects, such as the Dublin & Kingstown Railway, the projection of a railway from Liverpool to Leeds and the Manchester, Bolton, and Bury Railway in Lancashire.

Nimmo was also to leave his mark in South Wales on works such as Porthcawl Harbour, superintending its construction, with George Bush as principal assistant. He was also engaged by the harbour trustees at Aberystwyth to advise on proposed improvements, and his death in 1832 saw Bush being engaged as his replacement, on the recommendation of the Duke of Newcastle, then owner of the Hafod Estate in Aberystwyth.[61] Bush carried out his own survey, which largely agreed with Nimmo's plans, and a plan of the proposals, credited to 'Geo. Bush Engineer', was published in 1835.[62] Work began in 1836 with the building of a stone pier, the stone for which was quarried and transported along a specially built iron tramroad laid upon the embankment.[63] With the works completed, Aberystwyth became the third largest port in Wales in 1837, a situation that was not to last long, particularly in view of the works that engineers, including Bush, were to go on to. His projects in this field were to include dock works at Pembrey and Llanelli, as well as the Portland breakwater.[64] Bush had been involved in the construction of a tramroad for Aberystwyth and his first railway project was to come about through work at Llanelli.

On 10 January 1835, Bush submitted a report on the proposed Llanelly & Llandilo Railway to the Llanelly Railway & Dock (LRD) Committee.[65] The preamble covered the survey of a line from Llandybie to Llandilo: '… it was further agreed to extend the proposed line to the town of Llandilo-Fawr'. Bush proposed to commence the railway at the Llanelly floating dock, and to have a branch to communicate with the Llanelly Copper works. It required no major engineering works and there were to be few bridges of mention. Along the proposed line, stone was available that, '… will also afford good blocks for the purpose.' There would also be no inclination sufficiently steep to prohibit the use of locomotive engines with the steepest gradient being 1 in 108. Bush proposed that the main line should be formed, in the first instance, as a single line and, '... to be made a double line here-after, should the traffic require it'. There were to be a number of branch lines, including Genwen, Llwynhendy and Spitty. He estimated the cost of the main line (19 miles and 70 chains) plus thirteen branches (total 39 miles and 47 chains) at £149,500. A shortened version, with the main line to Spitty and three branches, was estimated at £16,500, while a junction railway between the present Llanelly Railway and the Carmarthenshire Tramroad (1 mile and 20 chains) was surveyed at £6,000.

The Llanelly Railway had been incorporated in 1828 to build a: 'Railway or Tramroad from Gelly Guile Farm… and for making and maintaining a Wet Dock at the terminus of the said Railway or Tramroad…'[66]

The wet or floating dock opened in 1833, along with the short 2-mile stretch of tramroad. D.S.M. Barrie describes the railway element as a horse-drawn tramroad, and also that the undertaking was not the first dock and railway venture in South Wales, as this distinction goes to the Duffryn Llynvi & Porthcawl Railway, incorporated some three years previously.[67] In 1835 the Llanelly Railway & Dock Co. obtained its Act of Parliament, authorising it, '… to make certain additional Railways or Tram-roads, and for other Purposes connected therewith.'[68] *The Cambrian* reported, in its 5 March 1836 edition, on the intended new railway from Llandilo to Llanelly:

> 'On Tuesday last, a grand dinner was given by the gentry of Llanelly, at the Thomas Arms Inn, to John Biddulph, jun., Esq., and others, the promoters of the intended new Railway to Llandilo, which is about to be commenced immediately, George Bush, Esq., the intelligent engineer, having arrived for that purpose… William Chambers, Esq., presided and was supported on the right by John Biddulph, jun., and Wm. Chambers, jun., Esqrs., and on the left by George Bush, Esq., and Thomas Morris, Esq., banker of Carmarthen. The Rev. E. Morris filled the Chair at the other end of the table…'

Aberystwyth Harbour. (SKJ photograph)

Left: Walter Coffin. Right: Thomas Powell. Deputy chairman and leading director, respectively, of the Taff Vale Railway during its formative years. (Courtesy of Cardiff Central Reference Library and Oxford House Industrial Archaeology Society)

There was a number of toasts, including one by the Revd E. Morris, who:

> '... expatiated on the immense advanges [sic] that would follow the expenditure of so large a sum (150,000*l*) as was required for the new Railway. "Mr. Bush's health" was next proposed and a high enlogium passed upon his talents as an engineer. He, in a neat speech, returned thanks for the flattering manner in which his health had been drunk...'[69]

John Biddulph of Ledbury, whose son, John Biddulph Jr, was involved in the railway because of his colliery interests, turned to Stephen Ballard to carry out surveys on the line. Ballard, the clerk turned engineer of the Hereford & Gloucester Canal, was to spend a week surveying the line at the end of 1836. He returned with his brother Philip to spend a further three weeks on railway business in May 1837.[70] The company's refusal to meet more than £50 of his £79 fee appears to have ended his relationship with the Llanelly Railway. Bush also appears to have ended his relationship with the Llanelly Railway before this, and does not appear on any published details on the Llanelly Railway after 1 January 1837. He is replaced by none other than George Stephenson, albeit in the capacity of 'consulting engineer'.[71] Bush was now working for the Taff Vale Railway.

Chapter 6 Notes

1 Dowlais Iron Company letterbooks, 1835 (1), letter 1109 20 November 1835, I K Brunel to Henry Scale.
2 John Steel had lost his leg in a boiler explosion and was killed in a second boiler explosion in France while Hugh Steel passed away in 1827 in unusual circumstances, Joseph Locke (1805-60) referring; 'to the melancholy death of poor Hugh Steel'.
3 Wood, Nicholas. *Practical Treatise on Rail-Roads.*
4 Barrie D.S.M. (1939, reprinted 1950, 1962 and 1969), p.6, *The Taff Vale Railway.* Oakwood Press, Lingfield.
5 University of Bristol, Special Collections, Brunel Office Diary DM 1758/1834, 7 October 1834.
6 Rolt, L.T.C., (1959), p.84. *Isambard Kingdom Brunel.* London: Longmans, Green & Co. 26 December 1835.
7 *Glamorgan, Monmouth and Brecon Gazette and Merthyr Guardian*, 12 October 1833.
8 *Glamorgan, Monmouth and Brecon Gazette and Merthyr Guardian*, 26 October 1833. Capt. Brown RN was also included on the committee. *The Cambrian*, 19 October 1833.
9 *The Cambrian*, 19 October 1833.
10 *Glamorgan, Monmouth and Brecon Gazette and Merthyr Guardian*, 26 October 1833.
11 *The Cambrian*, 19 October 1833, comment made by Mr Vivian.
12 Glamorgan, Monmouth & Brecon Gazette and Merthyr Guardian, 26 October 1833.
13 *The Cambrian*, 9 November 1833.
14 Mac Dermot, E.T. revised by C R Clinker. (1927, reprinted 1964, second impression 1972). *History of the Great Western Railway Volume 1.* London: Ian Allen. The scheme for the 'Bristol and London Railway' was issued on 7 May 1832.
15 MacDermot, E.T. (1972). p.9.
16 Williams, Archibald. (1925, reprinted 1972), p.10. *Brunel and After: The Romance of the Great Western Railway.* London: Patrick Stephens Ltd.
17 Ince, Laurence. (1993), p.58. *The South Wales Iron Industry 1750-1885*, Solihull: Ferric Publications.
18 Rolt, L.T.C., (1959). p.84. 26 December 1835.
19 Faraday had met with Guest on his visit to Wales in 1819 and Guest was to become a trustee, along with John Herschel, for the marriage settlement of Mary Ann Babbage (later Hollier).
20 Bessborough, Earl of. ed. (1950), pp.27-28. *Lady Charlotte Guest Extracts from her Journal 1833-1852.* London: John Murray.
21 Guest, Revel & John, Angela V. (1989), p.28. *John Lady Charlotte Guest A Biography of the Nineteenth Century.* London: George Weidenfeld & Nicolson. Extract from Elizabeth, Lady Holland to her son, Lord Ilchester, 1946.
22 Bessborough, Earl of, ed. (1950), p.35.
23 James, Brian Ll. ed. (1998), p.8. *G.T. Clark Scholar Ironmaster in the Victorian Age.* Cardiff: University of Wales Press. Further information on Frere's relationship with Brunel can be found in; Vaughan, Adrian. (1991, reprinted 1992). p.193. Isambard Kingdom Brunel, Engineering Knight-Errant. London: John Murray.
24 Lloyd, John. (1906). *The Early History of the Old South Wales Iron Works.* London: Bedford Press. 1834 was the last year in which Cyfarthfa totals exceeded that of Dowlais.
25 University of Bristol, Special Collections, Brunel Office Diary DM 1758/1834, 22 October 1834.

26 Glamorgan Record Office (GRO). Q/D/P/52. *Glamorgan, Monmouth and Brecon Gazette and Merthyr Guardian*, 15 October 1834, contains a notice of the Bill for the M&CR, but Colin Chapman informs me that the House of Lords Record Office has no record of the M&CR Bill or any proceedings.
27 Horsley, John Callcott. (1903). pp.176-78. *Recollections of a Royal Academician*. Dominie Sampson was a character in Sir Walter Scott's 'Guy Mannering'.
28 Brunel, Isambard, (1870, reprinted 1971). p.97. *The Life of Isambard Kingdom Brunel*. London: Longmans, Green & Co. (reprinted David & Charles, Newton Abbot).
29 Brunel, Isambard, (1870, reprinted 1971), p.97.
30 Horsley, John Callcott. (1903), pp. 178-79.
31 GRO. Dowlais Iron Company letterbook, 1835 (1), letter 1105-7, 9 October 1835, I.K Brunel to J.J. Guest.
32 Hadfield, Charles. (Second Edition 1967). p. 110. *The Canals of South Wales and the Border*. David & Charles: Newton Abbot (in conjunction with University of Wales Press Cardiff).
33 Hadfield, Charles. (1967). p.110.
34 *Glamorgan, Monmouth and Brecon Gazette and Merthyr Guardian*, 17 October 1835.
35 *Glamorgan, Monmouth and Brecon Gazette and Merthyr Guardian*, 17 October 1835.
36 *The Cambrian*, 24 October 1835.
37 *The Cambrian*, 31 October 1835.
38 Quotation by Joshua Bates, Partner in Baring Brothers. See Robert C.B. Miller, (2003). p.146, *railway.com: parallels between the early British railways and the ICT revolution*. London: IEA Research Monograph 57. Joshua Bates quotation from; David Kynaston, (1994). The City of London, A World of Its Own 1815–1890, Vol.1, Chatto & Windus: London.
39 *The Cambrian*, 26 September 1836.
40 Robert C.B. Miller, (2003). p.147.
41 Hadfield, Charles. (1967). p.110. Quoting from the Taff Vale Railway Proprietors' Minute Book, 12 October 1835.
42 RAIL TVR Provisional Committee minute, 23 October 1835.
43 *The Cambrian*, 21 November 1835. Also reported by the *Glamorgan, Monmouth and Brecon Gazette and Merthyr Guardian* for 21 November 1835.
44 GRO. Dowlais Iron Company letterbook, 1835 (1), letter 1109 20 November 1835, I.K Brunel to Henry Scale.
45 GRO. Dowlais Iron Company letterbook, 1835 (3), letter 846, 25 November 1835, Edward John Hutchins to T R Guest.
46 GRO. Dowlais Iron Company letterbook, 1835 (1), letter 1109, 20 November 1835, I.K. Brunel to Henry Scale.
47 University of Bristol, Special Collections, Private Letter Book 1, p.149, 4 September 1836, Brunel to Bush.
48 Jones, T.I. Jeffreys, ed. (1966), p.85. *Acts of Parliament Concerning Wales 1714-1901*. University of Wales Press: Cardiff.
49 University of Bristol, Special Collections, Letter Book 2, p.26.
50 Rolt, L T C. (1959), p.84, 26 December 1835.
51 University of Bristol, Special Collections, January 1836, Rickards to Brunel. The forthcoming second volume of *The Glamorganshire and Aberdare Canals* (Rowson, Stephen & Wright, Ian L.) will cover the Rickards tramroad and Blakemore's agreement with the Glamorganshire Canal.
52 Institution of Civil Engineers (ICE). Library Archives Ref 13a, First General Meeting of the TVR, 16 September 1836.
53 ICE Ref 13a, First General Meeting of the TVR, 16 September 1836.
54 University of Bristol, Special Collections, Letter Book 2, p.196, 5 December 1836.
55 University of Bristol, Special Collections, Letter Book 1, p.170. 14 October 1836 Brunel to Guest.

56 Title of the resultant Act of 1837 (1 Victoria).
57 Evidence on the Taff Vale Railway Bill, 26 April 1837.
58 Brunel, Isambard, (1870, reprinted 1971), p.94, *The Life of Isambard Kingdom Brunel*, Longmans, Green & Co.: London (reprinted David & Charles: Newton Abbot).
59 University of Bristol, Special Collections. Letter book 2, p.79, n.d.
60 Information from Mike Chrimes, Librarian, Insitution of Civil Engineers.
61 Aberystwyth Guide. (1848). Morgan: Aberystwyth.
62 National Library of Wales, Aberystwyth. Photo album 1262, PZ5463/95.
63 Troughton, William. (1997). p.9. *Aberystwyth Harbour An Illustrated History*. National Library of Wales: Aberystwyth.
64 Institution of Civil Engineers (ICE). Library Archives.
65 PRO. Ref: RAIL 377/104.
66 Jones, T. I. Jeffreys, ed. (1966), 9 Geo. IV (1828) xci (L. & P.) entry 557.
67 Barrie, D S M. revised by Baughan, Peter E. (1994), p. 210. *A Regional History of Great Britain, Vol. 12, South Wales*. David St John Thomas Publisher: Nairn.
68 Jones, T. I. Jeffreys, ed. (1966). 5-6 Wm IV xcvi (L. & P.) Llanelly Railway and Dock Co., entry 565.
69 *The Cambrian*, 5 March 1836, also quoted in *A Llanelli Chronicle*, (1984 & 1995) compiled by Gareth Hughes, Llanelli Town Council: Llanelli.
70 Bick, David E. (1979). pp.64-65. *Hereford & Gloucester Canal*. Pound House: Pound House.
71 ICE. Library archives 73a. Llanelly Railway and Dock Co., 1 January 1837.

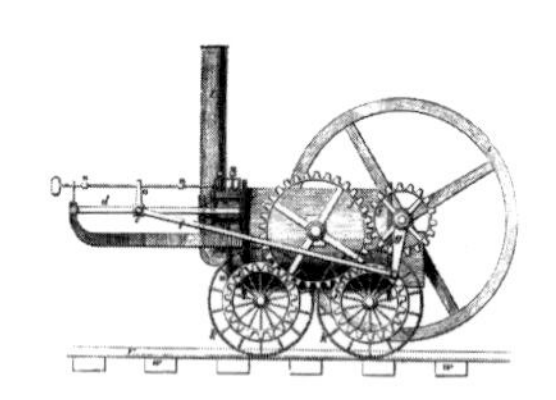

7

Engineer-In-Chief

'I Care Not However About It'

At this stage in his career, with projects such as the GWR coming to fruition as substantial commissions, Brunel needed to gather together a team of engineers and assistants, some of whom were to form the mainstay of his engineering office for many years. With so much engineering activity it was no wonder that Brunel allowed, or overlooked, a degree of autonomy on the TVR as far as the resident engineer, George Bush, was concerned. Angus Buchanan has written extensively on Brunel in a professional capacity, and on his relationships with his assistants and other engineers.[1] Others have drawn attention to his dictatorial nature, even making the claim that: 'He remains the only engineer whose relationship with his contractor has ever deteriorated to the point of premeditated physical violence.'[2] This refers to the famous 'Battle of Mickleton Tunnel', which occurred during the construction of the Oxford, Worcester & Wolverhampton Railway in 1851. While there would be no confrontations on this scale in South Wales, his relationship with Bush was far from ideal.

Brunel's role, or rather the role he exercised, did not improve the situation, particularly when he changed his mind about taking up the post of engineer on the TVR. When taking stock of his works on the eve of Boxing day in 1835, Brunel had said that he cared little for it as an engineering work, but Brunel did care who was 'engineer-in-chief.' There was also some ambiguity, at least to outsiders, regarding Brunel's role on the TVR, with contemporary accounts range from, 'Mr. Bush was joint engineer with Mr. Brunel'[3] to, '… George Bush, Esq, principal Engineer to the Taff Vale Railway Company'[4] It is interesting to note that the first modern history of the Taff Vale Railway does not mention Bush.[5] Bush was a well respected and confident engineer before he was employed on the TVR. He was also relatively young, being some five years younger than Brunel. In the 1841 census he is recorded as being thirty years of age, living in the Gabalfa Hamlet, part of the Llandaff parish, and relatively close to the Maindy works of the TVR in Cardiff.[6]

In a TVR Minute from 16 July 1836, Brunel was appointed engineer to the company at a salary of £400; however, it was also, 'Resolved that Mr Bush be appointed Resident Engineer to the Company at a salary of £700 pa.'[7] While this clearly indicates Brunel's appointment as company engineer, Bush's appointment, on a higher salary, is interesting and indicates the level of work that was expected.[8] Despite this, Brunel not only concerned himself with company matters but attempted to direct the work on a day-to-day basis. Some time later Bush believed he would become senior engineer when Brunel indicated that he wanted to reduce his commitment to the TVR; the board of Directors then sought to appoint Bush as 'principal engineer'. However, this expectation was short-lived as on 12 June 1837, Bush felt impelled to write to the deputy chairman, Walter Coffin:

> 'I have heard from Mr. Brunel that he had resumed the appointment of Principal Engineer to the Taff Vale Railway. I think you will perceive the propriety of my requesting that a minute should be entered in your Book, to be officially communicated to me that the negotiation opened with me by the Board of Directors to become their Principal Engineer is clarified.'[9]

And so the uneasy relationship between Bush and Brunel was to continue; Bush, appointed on a higher salary and subsequently asked to take on the role of principal engineer, was forced to work in Brunel's shadow. 'Nobody but the Chief was allowed a role of independent significance' is a phrase used by Angus Buchanan in his account of the team of engineers that worked for Brunel.[10] The TVR directors treated Bush with more respect, conveying directly to him rather than Brunel, important instructions such as,

> 'That Mr. Bush be instructed to proceed forthwith with the survey and preparing the property plans for the main Line and Branches… so that they may be ready by 25 December next, and that the survey and property plans for the Ely branch be proceeded with so as to be completed by 1 March 1838.'[11]

Before expanding on the working arrangements and relations under Brunel, there is some relevant information on two other members of the TVR engineering staff, the assistant engineers, Charles Bourns and Samuel Downing (1811-82). While laying out the line of the Taff Vale Railway, Bourns was to develop a formula for the calculation of the curvature of railways that was to earn him a Telford premium from the Institution of Civil Engineers, while Downing was to go on to become Professor of Engineering at Trinity College, Dublin.[12] In addition to assisting Bourns with his calculations, Downing worked on the largest engineering work on the Taff Vale Railway; the Goitre Coed Viaduct.[13]

Bourn's background was very similar to that of Bush in that he had worked on the Survey of Ireland and on the Dublin & Kingstown Railway, the early history of which was associated with Nimmo.[14] He joined the Institution of Civil Engineers in 1833 and resigned in 1851; at the time of joining the ICE he was based in London at 14 Bateman's Buildings, Soho Square. Bourns also had an interest in steamer vessel traffic in the Pool of London, carrying out a survey in 1835.[15] There was also an Irish connection with Downing, who was born in 1811 in Bagnelstown, Co. Carlow. After graduating from Trinity College he went on to further studies in

Edinburgh, then to work on engineering projects which included the construction of a 560ft-span timber viaduct connecting Portland Island with the mainland.[16] Downing was presumably known to Bush through work on the Portland breakwater and Bush was to offer him a position on the TVR on 28 April 1837: '… I can give you two guineas per week as proposed and will pay your travelling expenses to Cardiff…'[17]

Having completed the initial surveys, Brunel was to design the major works for the TVR in his Duke Street drawing office in Westminister (Downing only credits the design of the Goitre Coed Viaduct to Brunel). Leaving the day-to-day construction in the hands of George Bush, Brunel was to be so occupied with other projects that he was unable to attend the celebrations of the partial opening of the line from Cardiff to Navigation House (Abercynon) on 8 October 1840 (see Chapter 9). When he visited the works it would be in the form of flying visits made in his specially designed black *Britzska* coach. It was to provide not only his travelling office, containing drawing and engineering instruments, but his accommodation in the field with a couch-bed and some comforts, such as a stock of his ubiquitous Lopez cigars! Brunel's diaries show that every day of the week, including Saturdays, Sundays and Bank Holidays, particularly during the hectic period he found himself in during the latter months of 1834, was a working day. Most weekends he was either 'engaged on line' (GWR business); 'engaged on survey' (M&C); or travelling. The first day of November was a Bank Holiday (All Saints day) as well as a Saturday, but Brunel was up early on GWR business, for which he claimed half-a-day, and then, 'off by steam boat to Cardiff', for which he claimed a day on M&C business. The following Tuesday and Wednesday are also Bank Holidays (the Tuesday to celebrate the landing of King William III and the Wednesday to remember the 'Powder Plot'). Similarly, both days are fully occupied on GWR business.[18]

Through these flying visits, Brunel did not exactly neglect the Taff Vale Railway, at least as far as keeping an eye on the work of his assistants was concerned, and he made as many visits as he could to check progress. Visits would be made to inspect work in progress, the construction of the TVR's engineering depot at Maindy in Cardiff, as well as dealing with day-to-day company business and requests raised by directors. The preliminary surveys for the TVR led Brunel to make decisions about the route and the laying out of the line in a somewhat similar way to that adopted by the Glamorganshire Canal engineers some thirty-five years earlier (see chapter 8). In his 1836 estimate for the construction of the railway, Brunel refers to 'staithes and shipping places', which he estimated at £8,000. These, or rather their development into a full dock scheme, were part of a detailed report he wrote on 24 October 1836 on: '… the practicability of constructing Docks at the proposed termination of the Ely Branch of the Taff vale Railway.' This development of the TVR was not carried out, at least not under Brunel, and will be dealt with in Chapter 9.

The TVR was not the only railway project Brunel was involved with in South Wales as he was also dealing with the Gloucester & South Wales Railway proposals. In May 1836 he wrote to Mr Morris in Newport, his surveyor on the proposed Gloucester & South Wales Railway:

'I arrived at Newport just after you left the Inn, but could not stop. Press on the survey at Chepstow, Newport and Purton at the places indicated in order that I may mark a line there and have sections taken. Necessary also that you should immediately compile a map by connecting the surveys of different parishes and enclosing plans which you showed me so as to ascertain exactly what parts require filling up. I should wish to hear from you shortly as to your progress and to receive as soon as possible a tracing of the survey at Chepstow and Newport. Mr Bush and I will be with you very shortly.'[19]

Here we find the dilemma that Bush undoubtedly found himself in from time to time; he was employed by the TVR but reporting to Brunel and, therefore, liable to become involved in activities that were, strictly speaking, outside the TVR.

Further complications of this sort with the Gloucester & South Wales Railway (G&SWR) project stem from Guest's involvement. There are several letters from Brunel throughout 1836 addressed to James Kemp at the G&SWR office in Gloucester, mostly relating to his forthcoming report, one stressing that no surveys, '... could be made until the crops are off the ground.'[20] Brunel did, however, make the boundaries of his services crystal clear when the G&SWR directors tried to enlist his help in preparing and presenting Parliamentary evidence in support of their proposed line in September 1836, in a reference to the role of his Parliamentary Agent, St George Burke. St George Burke was to comment on their professional relationship during this period in a letter to Brunel's son, Isambard,

'I frequently accompanied him to the west of England, and into Gloucestershire and South Wales, when public meetings were held in support of the measures in which he was engaged, and I had occasion to observe the enormous popularity which he everywhere enjoyed. The moment he rose to address a meeting he was received with loud cheers, and he never failed to elicit applause at the end of his address, which was distinguished as much by simplicity of language and modesty of pretension as by accurate knowledge of his subject.'[21]

The financial panic of 1837, which marked the end of the little railway mania, was effectively to put the Gloucester & South Wales Railway on hold indefinitely. A revival of interest led to a separate undertaking in the shape of the later South Wales Railway (SWR) which gained its Act of Parliament in 1845. Work had been undertaken, however, on the G&SWR, and Brunel wanted to make sure that fees owed to his assistants on behalf of the SWR were not overlooked when he wrote to James Kemp early in 1837:

'I am somewhat surprised at your account of finances which you send me as when I last met the committee I understood that they had the means of so far meeting the expenses of the survey already completed with that which they argued should be continued to a certain extent – and all of which is I think within the amount I then estimated it at - However I take it for granted that immediate steps will be taken to enable you to discharge the balance, the more particularly that in a survey of this description the expenses consist almost exclusively of payments to surveyors - that which falls to me is so paltry as not to deserve the name of remuneration, certainly not profit and I cannot afford therefore to advance and to risk such payments - or even to wait long for them - And the poor operatives of course can but ill afford to wait. The Committee must therefore not look upon it in the light of some professional charges for which persons can afford to wait but more as to tradesmen's bills for which they have rendered themselves liable.'[22]

Although protecting their interests, Brunel is here referring to his surveyors and assistants as, '... the poor operatives...' who were in no doubt as to the working relationship expected by Brunel. In October 1836 he wrote to John Josiah Guest, stressing that he was keen to establish a relationship with the directors of the TVR that would prevent his assistants reporting directly on details of work in hand before he had the opportunity to analyse and report on it himself:

'As long as I enjoy the confidence of the Directors it is my wish, and I have particularly instructed Mr. Bush as my representatives fully to communicate information upon any thing connected with the works in progress but at the same time to avoid giving opinions upon subjects of any importance without first consulting me, as it must frequently happen that he may not be thoroughly informed upon my views and intentions, and if any reports be necessary, to refer to me ... It is impossible to write upon the most trifling subjects that which is to be made a document capable of being referred to (and reports to boards of directors always become such) without extreme caution, a caution quite inconsistent with that free and confidential communication which must exist between Directors and their professional advisers if the works are to proceed well.'

The 1870 biography on Brunel, by his son Isambard, devotes a chapter to 'Mr. Brunel's Professional Opinions and Practice' and there are extracts from several letters, such as one written in 1843: 'I have never accepted the position of joint-engineer; I have always refused to do so.'[23] How would he describe his position with Bush on the TVR, especially in light of further comments in letters dealing with his position on the role of consulting engineer:

'I never connect myself with an engineering work except as the Directing Engineer, who, under the Directors, has the sole responsibility and control of the engineering, and is therefore "The Engineer"; and I have always objected to the term "Consulting Engineer".' There are extracts in relation to contractors in which he explains his position, giving an example of a major litigation case between the GWR and a contractor which is one of, '...umpire between the Company and contractor.'[24]

This is an interesting comment in view of the problems, particularly payments, encountered by contractors working on the TVR. This biography uses extracts from several letters on the topic covered in his letter to John Guest above i.e. relations between the engineer and the directors and their 'interference' with assistant engineers.

Richard Beamish, who had worked with father and son on the Thames Tunnel and who was to write a biography of the older Brunel, also wrote on a topic somewhat removed from his engineering background. This was the *Psychonomy of the Hand*, published in 1865. Brunel, not surprisingly, was selected for examination and he goes into some detail on an analysis of his hand: '…an interesting illustration of an artistic development of the labour type.' Beamish includes the comment, 'Among the mental manifestations by which Brunel was distinguished, was an unhesitating and often overweening self-confidence.'[25]

The prompt settling of his account was another issue on which he would make his feelings crystal clear, as is demonstrated by the following letter to Joseph Ball of the TVR on 5 December 1836:

> 'I was very much surprised to find this morning that the balance of my account had not yet been remitted to the West of England Bank at Bristol. I am leaving town tomorrow for ten days and in making my arrangements for payments to be made during my absence I had calculated so fully upon the sum having already been paid to my account that I had drawn accordingly upon my Bristol Bankers and as I cannot now hear from you before I leave town I am much inconvenienced but I confess much more annoyed by this delay which under the circumstances might I think have been avoided. This account consists almost entirely of payments made by me for the Company long ago and it is most unprofitable and apparently a thankless office to be in ... I beg particularly to request that on Wednesday which I believe is your next board day the account be settled.'[26]

Delays and distractions from the real business of engineering was a major irritation to Brunel, including the addition of last minute alteration and additions. He was moved to write to the directors in October 1836, 'My attention has been called to an undertaking with Mr. Blakemore to make a branch to his works at Melin Griffith [the long established Melingriffith works to the north of Cardiff] which I had never heard of before. What is to be done in this matter?'

Blakemore was one of the objectors to the TVR who had to be bought off. A year later in September 1837, Brunel wrote to Mr Blakemore about carrying the line above the tail race of the Melingriffith works, between the narrow neck of land separating the tail race from the river. He informed Blakemore that he had stopped the works on the TVR for a few days to resolve this, and would consult with him as to the branch railway.[27]

Similarly, he wrote to Anthony Hill of the Plymouth ironworks in January 1837 about the question of the Limestone branch that Hill and the proprietors of the Penydarren and Dowlais works wanted to introduce to serve their works: '... the sooner we know you act for yourselves or through an Engineer... the more likely it is to be settled.' Despite this professional rebuke, Anthony Hill was to remain one of Brunel's friends and over twenty years later was to offer advice during Brunel's most testing time – the launch of the *Great Eastern*.

The uneasy relationship between Bush and Brunel was to come to a head in August 1838 when Brunel was asked by the deputy chairman of the TVR, Walter

Coffin, to clarify certain points about the permanent way. In order to reply, he needed data from Bush. The information did not come soon enough:

> 'My Dear Sir,
> How could you think of returning to Cardiff without seeing me - or rather how could you remain in Town two or three days after coming up expressly to meet me without seeing me or if any difficulty occurred, without writing a note to make an appointment. I am no doubt much engaged, but I waited anxiously and enquired day after day - not seeing you or hearing from you - a message or a note would have obtained an appointment at once. I have to make a report and for this I <u>must</u> have certain data –

The list of Curves	- as finally laid out
Do.	Gradients - Do.
Lengths of cuttings	- in Rock, gravel or materials requiring no bottom for ballasting
Do.	- in Clay or soft material
Do.	of embankment under 6' of Stone or Gravel
Do.	Soft Material
Do.	over 6' and under 15' Stone
Do.	Clay

> Particular description of all embankments above 15'.
>
> There is not a day to spare. I am afraid that the cause of the delay if any arises should not be known to the Directors or to say the least it was very thoughtless of you – and yet I know not how I am to receive the necessary information in time as I intend to be at Liverpool on Saturday and Dublin on Sunday – the only course that suggests itself to me is that you should send these data by post to Liverpool, addressed to me Adelphi Hotel, and a section by parcel ...to London on the chance of me being able to postpone my journey.
> Yours very truly IKB'[28]

Bush responded in a manner that Brunel was not used to receiving from his assistants. This prompted another letter:

> '... It is difficult in writing always to convey the feeling with which a remark is made and perhaps therefore I am mistaken in thinking your tone of reply is not quite what it ought to be... it is unnecessary to refer to it again but I must be able to make an observation such as I did a gentle reprimand if you will consider it so, without receiving such a dry reply, it is neither courteous nor kind – neither as my assistant or my friend...'[29]

Brunel knew that unless he was prepared to expend more time on the affairs of the TVR, which would be extremely difficult in view of other commitments such as the GWR, he could not afford to lose Bush. In the same letter he acknowledges that the

Launch of the PS Great Western *in Bristol on 19 July 1837. (Courtesy of Bristol Museum and Art Gallery)*

information sent is exactly right except for a little more detail! With a view to having a greater control over what is happening in South Wales he finishes the letter by asking: 'Have you printed forms for your assistants reports, I should wish to have them once a month in a general shape embracing the work of the month and describing the present state.'

Indeed the pressure on Brunel at this time must have enormous; he admits to Bush on 27 August 1838 that: '… Not being very well I went into the country immediately after the Bristol meeting and attended to no business of any sort which is the case of your letter of the 10th remaining unanswered…'[30] He goes on to confirm his confidence in Bush:

> 'In conducting the works of this Railway, I certainly look to you to carry out in detail the general principles which I may lay down or which are in accordant with the views you know me to entertain. If : in the present case, you entertain different views upon such important points as the permanent road and others referred to – and affecting the principle of construction, it cannot be as my representative that you report to the Directors, but do not let this cause any delay if such be the case, I shall write to you further on the subject.'

Bush did entertain different views on an important point – the gauge of the line, which again would put him in conflict with Brunel. A few days before Brunel wrote the above letter, Bush was part of a TVR deputation, paying a fact finding visit to the Stockton & Darlington Railway,

'Jeremiah Cairns, Newport, Mon. August 21st 1838.

I wrote to Mr. J. Pease to acquaint him that a deputation from the TVR would be in Darlington on Sunday or Monday with the purpose of looking over the S & DR and obtaining much information as possible to enable them to proceed with their large undertaking upon the best system but in the event of the absence of Mr Pease. I have advised them to call upon you and shall be obliged if you will give them your attention. The names of the Gentlemen are: Messrs Powell, Coffin & Hill their engineer Mr. Bush will accompany them.

To Samuel Barnard, Darlington, Secretary.'[31]

Brunel was interested to know what had been learnt on this visit; by 15 October he could contain himself no longer: 'I am very much surprised at not having heard nothing from you for so long – If you are returned from the north I should have wished to have known the result of your observations I also wish to know how the works are going on…'[32] The gauge of the TVR had now to be determined.

Brunel's choice of gauge for the GWR was the broad gauge of 7ft ¼in. The GWR was a line of communications built for speed and, despite some setbacks, Brunel's promise that the broad gauge offered high speed and safety was to be delivered. In December 1839 a test run from Paddington to Maidenhead achieved a top speed of 45mph.[33] In its 22 August 1841 edition, *The Cambrian* reported on an 'Experimental Trip on the GWR':

Brunel's Temple Meads station at Bristol, 1974. (SKJ photograph)

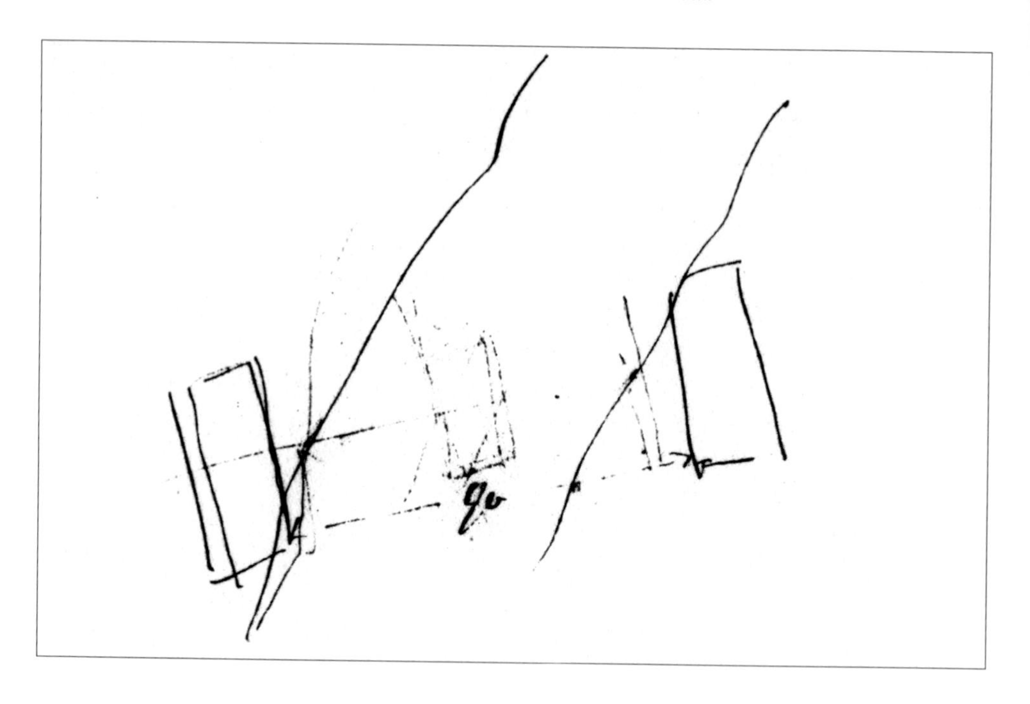

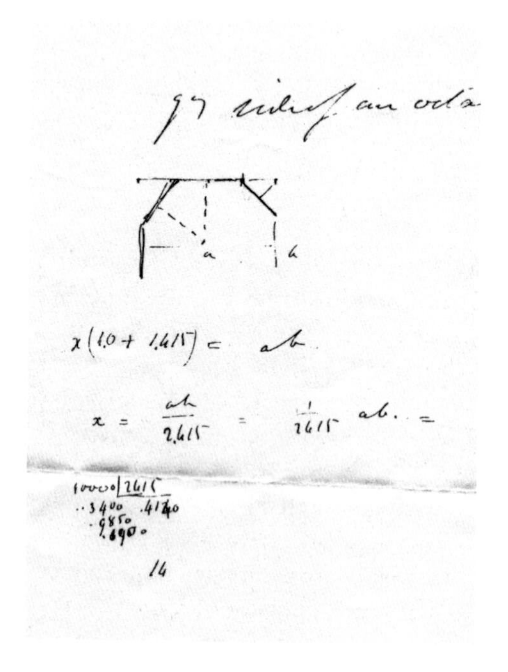

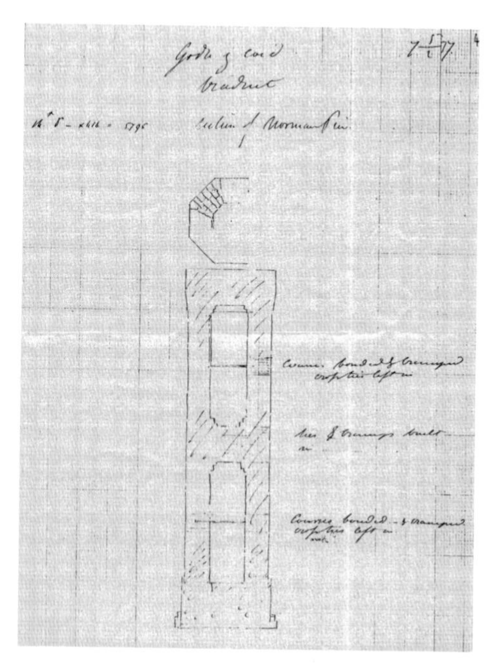

Brunel's sketches for Goitre Coed (Godre y Coed).
Top: *first sketch showing the problems of the river in relation to conventional piers.*
Bottom left: *the development of the octagonal shape.*
Bottom right: *final form of the octagonal masonry piers in the sketch book.*
(Courtesy of University of Bristol special collections)

> 'It is one of the railway on dits,[sic] that Mr. Brunel has wagered 1000L., that he will, after the completion of the Great Western, go from Bristol to London in two hours. If so we should prefer not travelling in the same carriage with him.'[34]

Such a report, acknowledged by *The Cambrian* to be only a rumour, has echoes of the £1,000 wager made on Trevithick's run. With regard to the experimental trip on the GWR, it is not known if the bet was paid; however, a week later *The Cambrian* reported that: '... the GWR to be opened Bath to Bristol on 31st inst... one of the engines went at a speed of 60 miles an hour, or a mile a minute.'[35]

The TVR was not the GWR, however, and the gauge was to be influenced, as Brunel was to recall later, by several factors such as the physical constraints of the terrain that the line was to pass through. Brunel recommended a gauge of 4ft 8½in or, rather, that he could see no reason for any other gauge. In his opinion this gauge was better suited for a mineral line, particularly in view of track curvature and future expansion, than his broad gauge of 7ft. In August 1838 he reported to the directors:

> 'As regards the gauge or width of the rails I see no reason in our case for deviating materially from the ordinary width of 4' 8½". The general gradients, the inclined planes and still more the nature and the immediate extent of the peculiar class of traffic to which the line must always be devoted not only render high speeds unnecessary but must almost prevent their being attempted, while the same causes operate to diminish any advantage that may be gained in reducing friction by increased diameter of carriage wheels. The curves also which the nature of the ground render unavoidable would be unfit for a wider gauge, but are not objectionable for moderate speeds with the ordinary gauge'.[36]

Bush was to return from his tour of railways and their operating practices in the north and, in November, submitted his report with a recommendation that the gauge for the railway should be 5ft-wide. The sub-committee set up to consider these issues accepted Bush's recommendation on 3 December 1838, as did a full meeting of the directors held on 18 January 1839. The decision was conveyed to Brunel, following which he advised that the railway should conform to the narrow gauge standard of 4ft 8½in. This was approved at a further meeting held on 3 April. Finally, in May 1839, Brunel's report recommended the narrow gauge 'standard' of 4ft 8½in to the TVR directors, rather than a 5ft-gauge for which it would have been difficult to obtain cheap rolling stock and locomotives. He felt it necessary to comment on future local industrial and commercial developments in support of his case:

> 'I cannot think that the day is very far distant when a direct communication will be made along the coast of South Wales with the interior and thereby to all the seaports in England. The district embraced by the Taff Vale Railway haying every possible facility must by that time have become a manufacturing district to a much greater extent than it now is, when indeed it may be said to produce and to export only the raw material; in such a situation when the utmost possible economy of transport must

be sought a direct junction between the Taff Vale Railway as a branch and such a main trunk would be essential. The Taff Vale would also in all probability be the means of communication from this line to the Port of Ely.

All these advantages would be obtained to the fullest extent if the suitability of gauge permitted a junction, but they would be lost or severely diminished if while the difference of gauges prevented a junction with the Taff Vale, other branch lines up the valley of Newport or elsewhere on to the Newport or Cardiff Docks had no such impediment. Now this main line will not be a 5' gauge, it will either be a 4' 8½"? or a 7' gauge, the latter you cannot have and therefore the former offers the only chance.'

Was it was a decision that he was later to regret? Possibly, when he was to give evidence to the gauge commission that had been appointed in 1845 to investigate the recommendation of a national railway gauge, but from the autumn of 1838 to early in 1839, Brunel was going through a crisis of confidence on the Great Western Railway. He was facing the outcome of external reports on his system: the gauge, the permanent way, the rolling stock and the works then under construction. At a crucial special meeting of the GWR on 9 January 1839, the Brunel camp managed to win the day. This was despite the external reports of the two engineers, Nicholas Wood and John Hawkshaw, whose reports providing the basis for an opposition proposed amendment that, '... the plans of construction pursued by Mr Brunel are injudicious, expensive and ineffectual for their professed objects, and therefore ought not to be proceeded in.'[37] The proposal was defeated and one of the GWR directors, George Henry Gibbs, was later to commit to his diary, commenting on the failed attempt of the northern proprietors group in the GWR, 'by our firmness we had secured their respect, kept them out of the direction, retained our engineer and preserved our gauge.'[38]

The surprising fact about this episode is that not only did Brunel put himself in such a position that would be exploited later, but that he failed to take notice of work that one of his assistant engineers was working on. Charles Bourns had published a 'Table of Gradients in the Minutes of Proceedings of the Institution of Civil Engineers' in 1838, and had based his work on the TVR.[39] He was to confirm the opinion of difficult terrain:

'Having been engaged in ranging the line of the Taff Vale Railway, which from the nature of the country presented circumstances of unusual difficulty, rendering it necessary to use curves constantly, and to vary their radii and flexure very frequently, my attention was particularly drawn to the principles upon which the practice of setting-out a line was founded, so as to preserve a proper continuity of curvature'.[40]

Basically, Bourns' proposition began from the principle that: 'It is not possible in practice to have a straight line of railway throughout, nor of one continuous degree of curvature; but it will consist of a system of curves and straight lines; the curves being not uncommonly of contrary flexure, and of different radii'.[41] Assisted by Samuel Downing in his investigations, Bourns came up with a method of calculating the setting out of railway curves in a plain and satisfactory manner as opposed to the inaccuracy of applying a square to the setting out of segmental curves, particularly those of short radii. His solution was to adopt segmental curves and he recommended the use of an offset staff as being a more theoretically correct and practical method when he presented his paper, 'On Setting-out Railway Curves'[42] to the Institution of Civil Engineers in May 1840. Using Bourns' proposition, the paper also included a table of offsets for curves of different radii, so that the whole practice of ranging could be performed with a chain, an offset staff and a few ranging poles.

If Bourns had come up with this proposition before Brunel had decided upon the gauge then he may not have made the decision he did. Giving evidence to the Gauge Commission, Brunel admitted that he had been mistaken when it came to calculating the amount of land needed to set out the line. Speaking of his reasons for adopting the narrow gauge on the Taff Vale Railway in 1838/39, he said:

> 'One of the reasons, I remember, was one which would not influence me now; but at that time I certainly assumed that the effects of curves was such, that the radius of the curve might be measured in units of the gauge, in which I have since found myself to have been mistaken'.[43]

Brunel therefore admits that the basis of his judgement was flawed, but does not refer to the source of the correct formula.

Accepting the obvious implications, there are two curious aspects to this episode. Why did Bush recommend a 5ft-gauge and why did the directors support his recommendation? There had been railways built to the 5ft-gauge, but very far and few between, notably the Eastern Counties Railway, which coincidentally, was incorporated just a month after the TVR. John Braithwaite was the engineer who was to adopt the gauge for good engineering reason, such as allowing on the locomotive, '... a little more space between the tubes...', but also to humour the directors of the Eastern Counties who wanted, '... a line on the grand scale of the *Great Western...*'[44] The latter probably also influenced the directors of the TVR, who did not accept Brunel's recommendation lightly; indeed he was questioned why they should not adopt the 5ft-gauge. The answer given above as part of his report was Brunel's reply. Despite the fact that the railway was being built, principally for traffic from Merthyr to Cardiff, the directors felt that if a broad gauge was good enough for their engineer to use on the Great Western, then it was good enough for the TVR. At only a few inches more that the 4ft 8½in-gauge, the 5ft-gauge was hardly a broad gauge, but that did not seem to matter. On this occasion, Guest does not appear to have displayed the strategic insight that was so characteristic of his approach to transport matters, when he supported the suggestion of the 5ft-gauge. It was left to Brunel to give this strategic vision by advising the directors that if the narrow gauge was to be adopted, it should be the standard 4ft 8½in-gauge and not the odd 5ft-gauge which was proposed by Bush.

As in the TVR and Eastern Counties examples above, the speed and safety advantages of the broad gauge seem to have found followers beyond the Great Western camp. In Ireland, a uniform gauge of 6ft 2in had been recommended in 1838, but

Samuel Downing, 1811-82. (Courtesy of Trinity College, Dublin)

this was changed in 1841 when the 5ft 3in-gauge was adopted, and this has remained the Irish standard to this day. The practical disadvantages of break of gauge in 1838 were not fully appreciated as the trunk lines under construction at the time had yet to form a coherent network. Many, like Hawkshaw, recognised the dangers posed by not adopting a uniform gauge, but the argument of having different gauges for main line and feeder was a popular one for the broad-gauge lobby, and one that Brunel himself was to promote. The disadvantages would be highlighted with the set piece disruptions staged for the Gauge Commission at Gloucester in 1845. In 1847, when Charles Frederick Cliffe was talking in his book about the proposed route of the South Wales Railway and the opportunity for iron and coal traffic, the difficulties posed by the break of gauge were not recognised, 'This kind of traffic stretches down to Cardiff, where a feeder will be met in the shape of the Taff Vale Railway.'[45]

Meanwhile, Bourns was to receive a Telford premium from the Institution of Civil Engineers for his work, noted as an: '… application of simple geometry, leading to practical results'.[46] The award went on to say that his method enabled engineers '… to set out curves of any radius and flexure, with a facility and precision not generally attained'.[47] In 1843 Bourns published The Principles and Practice of Land, Engineering, Trigonometrical, Subteraneous, and Marine Surveying[48], which included a section on setting out railway curves.

Bush did not confine himself solely to the TVR as he was also concerned with other railway projects. Wilkins, in his history of Merthyr, notes that Bush was engaged as engineer for the Brecon & Merthyr Railway from 1838 (the line surveyed by Edward Powell); 'A survey for Parliament was afterwards commenced at the instance of the late Sir John Guest., but was not completed due to the sudden death of Mr. Bush.'[49] He was also to carry out surveys in the Swansea valley, following on from an 1830 proposal and survey:

'Ten year later Mr. G. Bush, the engineer of the Taff Vale Railway, was employed to survey the valley, but opposition was again threatened unless the railway was made on the western side of the river, and as the estimated cost of 3 miles on the western side equalled the cost of the whole 16 miles on the eastern side, the undertaking was abandoned.'[50]

With the end of major engineering works on the TVR in sight, Bush began to reduce the number of staff employed in his department. On 7 May 1841, he was to write to Downing, whom he addressed as 'Dear Downing' (Bourns was usually addressed as 'My Dear Sir')

'A considerable reduction must take place in this department very shortly I think it therefore right to acquaint you that the Company will not require your Services beyond the next Quarter. I shall be most happy to bear my testimony to the ability and zeal you have display'd while upon these works and I need scarcely add that I shall be most glad to forward your views in any way you can point out if it should be within my power.'[51]

The half-yearly General Meeting of the TVR in February 1842 was to report that, 'The services of the engineering staff for the formation of the Railway being no longer required have been dispensed with.'[52] While Bush, as company engineer, would have been expected to remain in post through this period, he had already been 'dispensed with' by other circumstances. Bush died on 13 November 1841; he was thirty-one years of age and the cause of death was given as 'ulceration of the bowels'.[53] Despite his work in satisfactorily completing the main line, for which he had received glowing praise from the Inspector General of railways only months earlier as: '… testimony to the skilful manner in which Mr. Bush has grappled with the professional difficulties he has had to contend with…', he does not appear to have received the appropriate recognition for his work at the time of death and has been overlooked ever since.[54] His death is reported in the 'Deaths' section of the *Glamorgan, Monmouth and Brecon Gazette, Cardiff Advertiser, and Merthyr Guardian* in its 20 November 1841 edition:

'On the 13th inst., at the residence of his father, Lansdowne Place, Brighton, George Bush, Esq principal Engineer to the Taff Vale Railway Company. The death of this gentleman in the prime of his days, and the brightness of professional talents is deeply lamented, by the board of directors, who have so long experienced his superior abilities, and correct discharge of every official duty, while his amiable qualities in private life have endeared him to a large circle of sorrowing relations and friends'.[55]

There is no other comment in the newspaper in which many column inches are devoted to the news of the birth of the Prince of Wales. Despite his professional talents being, '… deeply lamented, by the board of directors…', the TVR board of directors did not respond appropriately and they do not even record Bush's demise in their meeting of 17 November 1841. However, they did resolve,'... that Mr Edward Bage be engaged as Resident Engineer to have the charge, under the General Superintendent, of the way and works and locomotive and stationary engines at a salary of £275 pa, payable quarterly.' It was also resolved, '... that a month's notice be given to all other persons employed on the Engineer's staff, except Mr William Page and Mr Fisher, that their services will be no longer required by the company.'[56]

The 'Mr Fisher' mentioned was George Fisher (1809-91), who was to play a major role in the future of the TVR. With the company facing straitened financial circumstance s, it was to act rather tardily in paying the salary arrears due to Bush at his time of death. In 1845, at a meeting of the board of directors, the secretary reported that the executors of Mr Bush were, '... proceeding with their action against the Company for the recovery of £769 1*s* 1*d* which they allege to be due to the estate of Mr. Bush, and that it was advised by the Company Counsel to pay the sum of £156 18*s* 11*d* with costs...', the latter considered to be the sum owing to Bush.[57]

It was to take almost three years for this to be resolved, as on 11 January 1848, 'Read a letter from Messrs Hunt dated 8th inst [January 1848], that the Plaintiffs in the Action "Bush versus T.V.R.C." have replied "satisfaction" and abandoned further proceedings. The taxed costs outstanding £14 11*s* 0*d*. Messrs asks for a remittance.'[58]

Downing went on to take up the post of Assistant Professor at Trinity College in 1846, becoming Professor of Practical Engineering in 1852. His lectures were, '... illustrated by diagrams and models of the most recent celebrated works', and surviving accounts of Downing's teaching methods indicate that one of his teaching aids was a large illustration of the Goitre Coed Viaduct.[59] The viaduct was to be the subject of a paper in the Transactions of the Institution of Civil Engineers of Ireland in 1851 (see chapter 8).

Brunel was to maintain his links with the company, albeit now at somewhat of an arm's length. In 1843 he was asked to report on the state of the railway, the TVR having taken out a maintenance contract in December 1841. There were, in fact, three elements to his report; the contract itself, the state of the permanent way with reference to the contract entered into and what extra works should be undertaken.[60] Brunel reported that the general state of the works did not indicate any neglect of the terms of the contract except that the drainage of ballast was decidedly defective; that there appeared to be neglect in replacing defective materials and that, '... the original sleepers do not appear to have been so good as they ought to have been...' – so much for the 'bargain' of using old ship's timbers! Brunel accepted that, '... the line may be said to be in tolerably working order but will shortly need considerable repairs... £6000 to £8000 would be sufficient'. How much the TVR actually committed itself with regard to repairs is debatable considering its financial position, but extra work, such as a start on doubling the main line, was to be undertaken over the next few years. Cost saving was probably the main reason for the maintenance contract in the first place as the TVR may have thought it prudent to enter into such a contract following the death of Bush, although it was not an unusual arrangement for newly constructed railways.

In the period following Bush there were several appointments, a notable recruit being the George Fisher already mentioned, although his appointment appears to have been one of the last appointments made by Bush, as according to Wilkins he joined the TVR in 1840.[61] Fisher had been previously employed on the harbour works at Aberystwyth and so, presumably, was known to Bush.[62] In the period following Bush's death, Fisher was appointed as superintendent but resigned from his position on the TVR in September 1843. His replacement was Edward Highton, but Fisher was still involved as a contractor (see Chapter 9). With Highton's departure at the end of 1845, Fisher returned to permanent employment with the TVR, and stayed with the company until his death in 1891. At the time of his death he was deputy chairman of the TVR and was, successively, to hold the posts of superintendent, engineer and general superintendent, engineer and general manager before becoming resident director (i.e. managing director) in 1883.[63]

In 1846 Brunel was requested by the board of directors to tender his professional assistance to the company in passing the Bill through Parliament for the extensions they proposed to apply for in ensuing session.[64] Just over a year later, the secretary was again requested to write to Brunel, this time in relation to the South Wales Railway, to press him 'immediately' on setting out and levelling the line of the SWR where it crosses the TVR: '... that the building of the Bridge may be commenced without further delay.'[65]

The SWR was now Brunel's priority in South Wales, as far as railway engineering was concerned, but there was still work to be done. On 5 January 1846, he reported on the, '... New Work and Contracts now in execution for the Doubling of the Taff Vale Railway.'[66] The first part of his report states that the first contract is by Hugh Allen to form an embankment for the East branch line to the Bute Dock, which involved taking earth from the field at the Cardiff station. Brunel's role as principal engineer for the Taff Vale Railway was now virtually complete; the original main line of the Taff Vale Railway, with the exception of the Ely branch and the northern branches, had been completed, along with the branches into the Rhondda and Llancaiach, and the TVR now enjoyed access to the Bute Dock, although it was not entirely to their satisfaction and was to lead to problems in the not too distant future.

TVR letterhead device. (SKJ collection)

Chapter 7 Notes

1. Buchanan, R.A., (2002). *Brunel The Life and Times of Isambard Kingdom Brunel.* London: Hambledon and London, see Chapter 10 on 'The Professional Man' pp.153-71. See also Kentley, Eric, Hudson, Angie and Peto, James. (2000) Isambard Kingdom Brunel:Recent Works. London: Design Museum. Chapter 1 Working for the Chief. Buchanan, R A.
2. Barnes, Martin. (2000). p.138. *Civil engineering management in the Industrial Revolution.* Civil Engineering, 138, August 2000, pp.135-144, Paper 12109.
3. Cliffe, Charles Frederick. (1847 1st ed). p.63. *The Book of South Wales.* London: Hamilton Adams & Co.
4. *Glamorgan, Monmouth and Brecon Gazette, Cardiff Advertiser and Merthyr Guardian,* 20 November 1841, obituary notice for George Bush (13 November 1841).
5. Barrie, D.S.M., (1969 ed. originally pub.1939). *The Taff Vale Railway.* London: The Oakwood Press.
6. 1841 Census return. Llandaff schedule 136, entry 4.
7. PRO. RAIL. TVR Minutes 16 July 1836.
8. Colin Chapman suggests that the role of Brunel and Bush could be seen, in modern terms, as that of Consulting Engineer and Resident Engineer.
9. PRO. RAIL 684/64. p.4. George Bush's Letter Book.
10. Kentley, Eric, Hudson, Angie and Peto, James. (2000) *Isambard Kingdom Brunel: Recent Works.* London: Design Museum. Chapter 1 Working for the Chief. Buchanan, R.A.
11. PRO. RAIL.TVR Minutes 25 October 1837.
12. Cox, R.C., (1993). pp.24-39. *Engineering at Trinity.* (incorporating a Record of the School of Engineering), Dublin: School of Engineering, Trinity College Dublin. Chap 1 deals with 'Foundation and Early Years' (pp.7-34) and Chapter 2: 'The Downing Era' (pp.35-58). See 13 for Bourns.
13. Downing, Samuel. (1851). pp.23-48. *Description of the Curved Viaduct at Goed-re-Coed near Quakers Yard, Taff Vale Railway.* Trans. Institution of Civil Engineers of Ireland. Vol.4, Part 1.
14. Nimmo was replaced on the Dublin & Kingstown Railway by Charles Blacker Vignoles.
15. Institution of Civil Engineers (ICE) library references, Port of London, ref. O.C/173, 1835. Retaining walls at Kingstown, near Dublin, ref. O.C/184, 1835. Table of gradients, ref. O.C/281 [n.d.]. Trans. Institution of Civil Engineers. (1838). p.49. Table of gradients.[nl] Bourns, C,[nl] Minutes of the proceedings, volume 1, session 1. On Setting Out Railway Curves.[nl] Bourns, C,[nl] Mins of the Proceedings, Vol 1, session 1840, p. 56. (Taff Vale,Wales). *On Setting-Out Railway Curves.*[nl] Bourns, C,[nl] Trans Institution of Civil Engineers, pp. 295-303. Vol.3, part 4, session 1841.
16. Cox, R.C. (1993). pp.24-39. *Engineering at Trinity.*
17. PRO. RAIL 684/64, p.2, George Bush's Letter Book.
18. University of Bristol, Special Collections, IKB Office Diary DM 1758/1834, Oct. & Nov. 1834.
19. University of Bristol, Special Collections, IKB Letter Book.
20. University of Bristol, Special Collections, IKB Letter Book, p.133, to J. Kemp, 29 July 1836.
21. Brunel, Isambard. (1870 reprinted 1971). pp.75-78. *The Life of Isambard Kingdom Brunel.* Newton Abbot: D&C reprint. *Reminiscences of Mr. Brunel 1833-1835,* letter from St George Burke QC.

22 University of Bristol, Special Collections, IKB Letter Book. p.220, letter to J. Kemp, 20 January 1837. I am grateful to Colin Chapman for pointing out an action by Brunel to recover fees from the G & SWR committee as late as September 1843 (reported in *The Railway Times*, 2 September 1843).
23 Brunel, Isambard. (1870, reprinted 1971). pp.476-77. *The Life of Isambard Kingdom Brunel.* Newton Abbot: D&C reprint. On the position of Joint Engineer. 16 October 1843.
24 Brunel, Isambard. (1870, reprinted 1971). pp474-83. *The Life of Isambard Kingdom Brunel.* Newton Abbot: D&C reprint. The litigation was Ranger v. the GWR.
25 Beamish, Richard, (1865), p.55-56, *The Psychonomy of the Hand*, Frederick Pitman, London.
26 University of Bristol, Special Collections IKB Letter Book, p.1976, to J Ball, 5 December 1836.
27 University of Bristol, Special Collections IKB Letter Book, p.254, to Blakemore, September 1837.
28 University of Bristol, Special Collections IKB Letter Book, p.316, to G Bush, 1 August 1838.
29 University of Bristol, Special Collections IKB Letter Book, p.318, to G Bush, 4 August 1838.
30 University of Bristol, Special Collections IKB Letter Book, p.320, to G Bush, 27 August 1838.
31 PRO. RAIL 667/1099 Stockton & Darlington Railway, 21 August 1838. Jeremiah Cairns had played a significant role in the development of the SDR and was now living in Newport (see Chapter 4).
As well as being Thomas Meynell's agent, Jeremiah Cairns had also subscribed to the SDR in 1821 and had sought tenders from Dowlais ironworks on the supply of rails when the line was being built.
32 University of Bristol, Special Collections IKB Letter Book, p.333, to G. Bush, 15 October 1838.
33 MacDermot, E T, revised by Clinker, C R. (1964). p.45. *History of the Great Western Railway*, Vol.1. London: Ian Allen.
34 *The Cambrian*, 22 August 1841 (Sat) Experimental Trip on the Great Western Railway
35 PRO RAIL 1014/30/6 Report of I.K. Brunel to Directors of Taff Vale Railway on the selection of a gauge pf 4ft 8½in, 23 April 1839.
36 *The Cambrian*, 29 August 1841.
37 MacDermot, E T, revised by Clinker, C.R. (1964). p.47. *History of the Great Western Railway*, Vol.1. London: Ian Allen.
38 Gibbs, George Henry. ed by Simmons, Jack. (1971). p.67. *The Birth of the Great Western Railway, Extracts from the diary and correspondence of George Henry Gibbs*. Bath: Adams & Dart, Bath.
39 Min. Proc. Institution of Civil Engineers 1838.
40 Trans. Institution of Civil Engineers. Vol iii, Part 4.
41 Trans. Institution of Civil Engineers. Vol iii, Part 4.
42 Trans. Institution of Civil Engineers. Vol iii, Part 4.
43 Brunel, Isambard. (1870 reprinted 1971). p.104. *The Life of Isambard Kingdom Brunel.* Newton Abbot: D&C reprint. The Broad Gauge.
44 Day, Lance, (1985). p.35. *Broad Gauge*. London: Science Museum.
45 Cliffe, Charles Frederick. (1847 1st ed). pxii.
46 Trans. Institution of Civil Engineers. Vol iii, Part 4.
47 Trans. Institution of Civil Engineers. Vol iii, Part 4.
48 Bourns, Charles. (1843). *The Principles and Practice of Land, Engineering, Trigonometrical, Subteraneous, and Marine Surveying*. With an appendix. London: John Ollivier.
49 Wilkins, Charles. (1908, 2nd ed.). [p.194]. *The History of Merthyr.* Merthyr Tydfil: Williams.

50 Jones, W.H. (1922). *History of the Port of Swansea*, Carmarthen: Spurrell. Further references on the role played by Starling Benson in appointing Bush to survey the proposed line can be found in Bayliffe, Dorothy M. and Harding, Joan N. (1996), *Starling Benson of Swansea*. Cowbridge and Bridgend: D. Brown & Sons Limited.
51 PRO. RAI.L 684/64. p.167. George Bush's Letter Book.
52 *Glamorgan, Monmouth and Brecon Gazette, Cardiff Advertiser and Merthyr Guardian*, 26 February 1842 on the half yearly General Meeting of the TVR.
53 He died at the home of his father; Richard Bush Jr, 16 Lansdowne Place, Hove.
54 *Glamorgan, Monmouth and Brecon Gazette and Merthyr Guardian*, 24 April 1841.
55 *Glamorgan, Monmouth and Brecon Gazette, Cardiff Advertiser and Merthyr Guardian*, 20 November 1841. Obituary notice for George Bush (13 November 1841). Not reported in *The Cambrian* until the following week (27 November) similar report to the above.
56 PRO. RAIL. TVR Minutes 17 November 1841.
57 GRO. RAIL 684-2 1908, 19 November 1845.
58 GRO. RAIL 684-2 3116, 11 January 1848.
59 Cox, R.C. (1992). Ron Cox of Trinity College informs me that Downing used an enlarged copy from his paper of the elevation of the viaduct in his presentation to students, ie. large enough to be seen from the back of the lecture room (information supplied 31 July 2003).
60 University of Bristol, Special Collections IKB Letter Book 2, p.197. To W Coffin, 31 January 1843.
61 Wilkins, Charles. (1908 2nd ed). p.194.
62 Chapman, Colin. (1992). p.163. *Welsh Railways Archive*, Vol 1, No.6. *A Biography of George Fisher.*
63 Chapman, Colin. (1997). p.143. *The Nelson and Ynysybwl Branches of the Taff Vale Railway*. Oxford: The Oakwood Press.
64 PRO. RAIL 684-82 1962, 7 January 1846.
65 PRO. RAIL 684-82 2720, 10 February 1846.
66 PRO. RAIL 684-856, 5 January 1846.

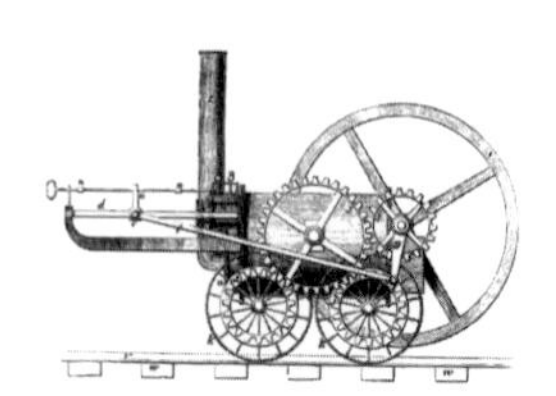

8

Engineering Works

'The Viaduct Looked as Beautiful as Ever...'

For both the GWR and TVR, in terms of their original phases of construction, 1838 marked a year in which major engineering projects were in progress. It was also a year of decisions that were to have a profound effect on the future direction of both railways. On the GWR there were a number of major works, such as the Maidenhead Bridge, while on the TVR, the Goitre Coed Viaduct was the most important.

In drawing parallels with Maidenhead and Goitre Coed, it is interesting to note that in 1838, the year that the centering was eased on the Maidenhead Bridge, the engineer Thomas Macdougall Smith was presenting a paper to the Institution of Civil Engineers.[1] The paper described the failure of William Edwards' spans across the Taff at Pontypridd and the final successful 140ft-single-span on the site.[2] The failure of the first two spans was caused by pressure on the haunches which forced up the crown of the bridge. William Edwards' answer, achieved at the third attempt, was to add weight to the crown and lighten the haunches by piercing the spandrels with three cylindrical holes. It is believed that John, later Sir John, Hawkshaw was present at this meeting of the Institution of Civil Engineers in 1838, which may have influenced him to comment in his report on Maidenhead Bridge that: '... more weight was required on the crown'. Hawkshaw was later to be involved in a number of engineering works that followed on from original projects of Brunel's, such as the Clifton Suspension Bridge. In 1838, however, he was to criticise the construction of Maidenhead Bridge and condemn the broad gauge in a report that was commissioned by the GWR.[3] With regards to Maidenhead Bridge, Brunel did not respond to the comment but it is known that he had reduced the weight of the bridge haunches in that, '... the longitudinal walls were used "as ribs" in the spandrels which are not solid.'[4]

Brunel would have had an opportunity to examine the 'Pont y tŷ pridd' during his visits to the Merthyr ironworks or the nearby Brown Lenox Chainworks before Smith's paper was presented. This practical examination, and a thorough grounding in mathematical analysis from his father, was to form his judgement in this area, and when problems arose he held fast to his calculations;

'Brunel did not attempt to load the crown of the Maidenhead arch as suggested by Hawkshaw but in the knowledge that he could rely on the support of the centering, chose to wait for the weight of the bridge to consolidate the abutments (in case of any movement there) and for the mortar to set. There would be no trouble if he could but obtain his calculated horizontal thrust from the consolidated arch and abutments.'[5]

Work had begun on Maidenhead a year or so before the major work on the TVR, the Goitre Coed Viaduct. This was contract No.1: 'Being the construction of a Stone Bridge of 90 feet in height, across the River Taff at Goche-y-coed [sic]...'[6] The contract also included the excavation and formation of the tunnel and cutting at Goitre Coed, as well as the formation of the embankment, culverts and other masonry work. Tenders for these contracts, and No.3 (Ynyscoi tunnel near Taffs Well) and No.4 (bridge over river Taff at Melin Griffith) were being sought by advertisement in March 1837.[7] The award of the contracts for Merthyr embankment and Ysgybon Newydd incline was recommended to William Lewis: '... he has facilities for obtaining cinders from Messrs. Hills works – and on that account can execute the work cheaper than other contractors...'. Contract No.3 was awarded to Evan Evans to commence on Monday 25 September 1837.[8] The closing dates for receipt of tenders for contract Nos 1, 3 and 4 were extended to 26 April, presumably as No.2, Newbridge Viaduct, was being awarded (on Brunel's recollection this was to the contractor Jones, but the notice to commence for contract No.2 was actually issued to John Edmunds Esq.[9]). Later in the year, tenders were also sought for Nos 5 and 6 for timber bridges at Taffs Well and Pentyrch respectively.[10]

Pont-y-tŷ-pridd, William Edwards' bridge. The name originally applied to the hamlet where the bridge was situated, with the shortened form of Pontypridd eventually taking preference over Newbridge by the late 1850s as the place name. (SKJ collection)

William Edwards, 1719-1789, the architect and builder of Pont-y-tŷ-pridd. (SKJ collection)

While contractors were found for the major works, Brunel was having problems finding contractors for the main line, and the story goes that he approached Robert Stephenson and sought advice from him regarding suitable contractors. Stephenson, who had previously tested the contracting skills of John Calvert (1812-90), sent for him, apparently suggesting that he should, '… tender for a certain crooked line down in Wales.' When the nature of the work was outlined to Calvert, he is reported to have replied, 'No, I won't undertake it, why, they cannot even speak English down there'.[11] Calvert was eventually persuaded to change his mind by the combined efforts of Stephenson and Brunel, and he undertook the contract, constructing the railway between Llandaff and Merthyr. The experience appears to have had an impact on Calvert, who must have warmed to Wales as he stayed in the locality after the contract. Indeed, he never returned to Yorkshire, but was to become one of the pioneers of the Rhondda coal industry (see Chapter 11).

There appears to be little documentary evidence covering the progress of the works, with the exception of Samuel Downing's paper on its construction on the Goitre Coed Viaduct.[12] That the works should have caused some interest, at least to Cardiffians, is picked up in a fictional work by the novelist Jack Jones who, in River out of Eden, refers to a comparison with the rates paid to the navvies employed on the Bute Dock and the TVR (see Chapter 9).[13] There is also a reference to the hero of that book, Dan Regan, taking the role of head ganger for a contract for which he had '…an interview at Abercynon with the railway engineer.'[14] This brief fictional account relates to Regan and his 120 navvies working on the: '… stiffest cuttings that have to be opened up for the railway.'

Brunel issued his own conditions of contract and notices to commence through the TVR office, based on his experience of dealing with contractors, the profession being too young to have developed any standard forms. It has been argued that

Above: *Newbridge Viaduct, Pontypridd. The Newbridge Viaduct was the first major work on the TVR for which a contract was granted. Here, the first stone of the TVR was laid in a ceremony performed by Lady Charlotte Guest (wife of Sir J.J. Guest of Dowlais) on 16 August 1837. The original viaduct was widened to accommodate a double line in 1862 by building a viaduct with smaller spans alongside. The middle pier of this viaduct can be seen through Brunel's skew arch. (SKJ collection)*

Left: *Site of tunnel (known as Penlock's Tunnel) at Quakers Yard. Note the tunnel wall still in situ. (SKJ photograph)*

Brunel's approach led to many disputed claims as his form of contract was modified with each successive commission to erode the rights of the contractor. Consequently, some claims were taken up through the courts. The most famous of these concerned the contractor McIntosh, and was over problems encountered on two sections of the GWR. It was eventually decided in McIntosh's favour by the Court of Chancery in 1865 – some twenty-nine years later! There was no cause for celebration as both McIntosh and Brunel were dead by this time and, furthermore, the judge was highly critical of Brunel's management practices.[15]

Victorian guidebooks on localities such as Maidenhead related to the difficulties and risks that had attended the construction of Maidenhead Bridge. But Brunel was not in the business of taking risks at Maidenhead, Goitre Coed or anywhere else for that matter, as he employed mathematical calculations to determine the forces at work, what we would now term 'finite element analysis'. Today, finite element analysis is an effective numerical method used to model a wide range of physical phenomena. In Brunel's case, it was used to analyse every structural element, down to an individual stone in a bridge, so that the forces acting on it could be individually calculated. Brunel was able to calculate the force acting on each element up to five significant figures, and ensure that the forces between each stone were acting in compression. As a building material, stone has very little strength in tension but is exceedingly strong in compression. Another modern description of Brunel's direct and safe approach to arch design would be 'statically admissible'.[16]

While the works of the line appear to have been little noticed in terms of written records, the Goitre Coed Viaduct had attracted the attention of the artist Penry Williams as an artistic subject. The construction of the viaduct, a graceful, lofty and striking design, was an object of great interest. In a parallel situation to what was happening at almost the same time over 100 miles away on Brunel's Maidenhead Bridge, visitors to the Goitre Coed site were to express strong opinions concerning its ultimate stability.

At Maidenhead, Brunel had designed a brick viaduct with two flat, semi-elliptical main arches, each 128ft-wide and only 24½ft-high – the largest and flattest brick arches that had ever been built. The critics predicted that the bridge would collapse as soon as the centerings were eased, which incidentally were of a similar type to those used at Goitre Coed i.e. of the improved fan type. This failure of the critics' prediction is today well known, and George Henry Gibbs, one of the GWR directors, mentions in his diary (21 June 1838) the effect that the critics were causing:

> 'The most ridiculous reports were in circulation today about the bridge and we were so pestered with inquiries that it became necessary to prepare some sort of report to tranquillise the shareholders…'[17]

Less well known are the criticisms levelled at Goitre Coed. The assistant engineer on the Goitre Coed Viaduct was Samuel Downing, who dismissed one of the objections as:

*'Maidenhead Viaduct' by J.C. Bourne. (*From The History and Description of the Great Western Railway*, 1846)*

> '... not deserving of any answer, that namely which founded upon the circumstance of the curve of the plan being concave to the direction of the force of the river; had it been curved convex to the current, it would then have had (said they) the strength derived from its arched form'.[18]

Another objection was taken more seriously:

> '... that the thrusts on the top of any pier, arising from the two half arches which it supported, could not possibly equilibrate, since their directions must form an angle, equal to that at the centre of the curve of the bridge, subtended by a chord whose length was the distance from centre to centre of the piers'.

It is not clear who came up with the calculation, but Downing states:

> 'The answer to this was derived from a computation which gave the actual amount of the unbalanced resultant of the horizontal thrusts as nearly 5 tons; now this acting with a leverage of 42 feet (the greatest height of the pier namely), gives its moment as 210 tons. The total weight upon the base of the loftiest of the piers was about 1400 tons, and this multiplied by 7 feet, its leverage (half the diameter of the pier), gives for the moment of the stability 9,800 tons – a result which shows the groundless nature of the objection, the forces tending to stability, being to the unbalanced thrusts as 47 to 1'.[19]

The comments levelled at Maidenhead, as far as official reports were concerned, proved to be groundless, although a young engineer named John (later Sir John) Fowler thought that the problems were serious. In his opinion, disaster was only avoided by the centering used, which he described as: '… a very excellent and scientific one, the whole of the timbers being in a state of thrust.'[20] Accounts such as MacDermot's relate the concern made when the centering under one of the arches was eased prematurely before the mortar had set. When the damage was repaired, Brunel was to leave the centering in place to placate the critics, knowing that the centering was actually eased away from the stonework and played no structural role. Indeed, it was to blow down in the following winter with no effect on the stability of the bridge! It is not known if a similar move was suggested for Goitre Coed but Penry Williams shows that the centering is still in place under the tramway arch after the viaduct was opened for traffic. It is more likely that this, the last piece of centering, had not been taken down when he sketched the viaduct.

Downing was later to give an engineer's examination of the course and construction of the Taff Vale Railway and suggested that reference to the ordnance map of the river and its tributaries would, '… put us in the position of the engineer whose duty it was to determine these circumstances'.[21]

He went on to add that the canal, as an earlier undertaking, had taken first choice; '… this preoccupation of the best locality has put the rival railway to very considerable expenses'.[22] Despite this comment, Downing went on to state that the ultimate course was the best in terms of the overall inclination of the line and in securing access to tributary valleys for coal exploitation. There was no conflict, however, on the starting points for the canal and railway, which were the west and east side of the Taff

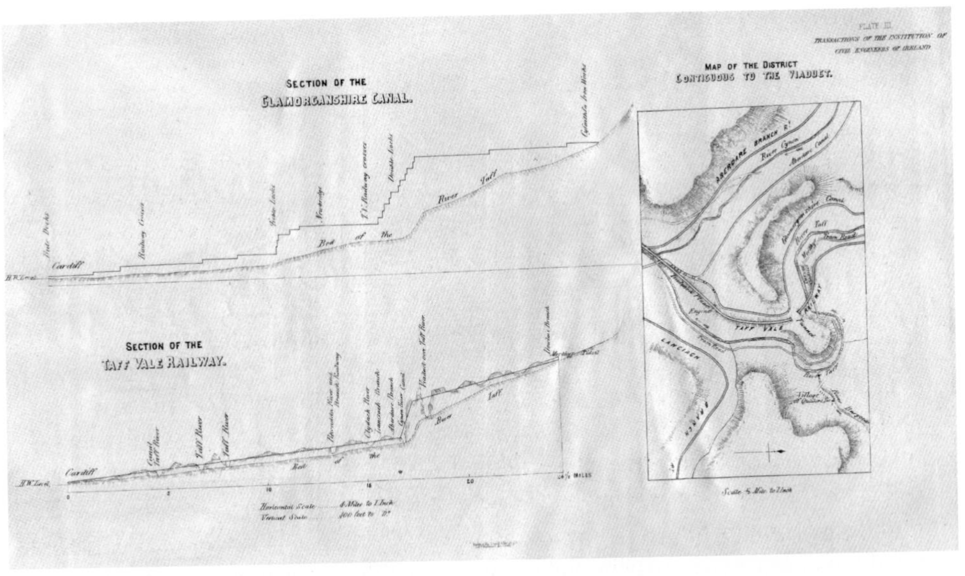

'Goed-re-Coed Viaduct, gradients of railway and canal' by Samuel Downing. (From Trans. Institution of Civil Engineers of Ireland, 1851)

at Merthyr respectively, because of vested interests. The canal's main financial investors were the Crawshays, whose Cyfarthfa works were situated on the west side. The railway was powerfully supported by Guest of Dowlais and the other ironmasters on the eastern side of the river. Here the railway's Merthyr terminus would be built, with the line proceeding down the valley, on the left or eastern side, for around 8 miles, that is until it reaches the point where Goitre Coed Viaduct crosses over the Taff.

It was below this point that Brunel decided to concentrate the principal gradient of the railway, the line rising from the shipping place at Cardiff to the terminus at Merthyr. In this respect he was closely following the engineering solution arrived at some forty years previously by the canal engineers. At this point the section of the canal runs very close to that of the railway, and the solution in overcoming a nearly equal fall was to employ eleven locks. On the railway, however, an inclined plane was

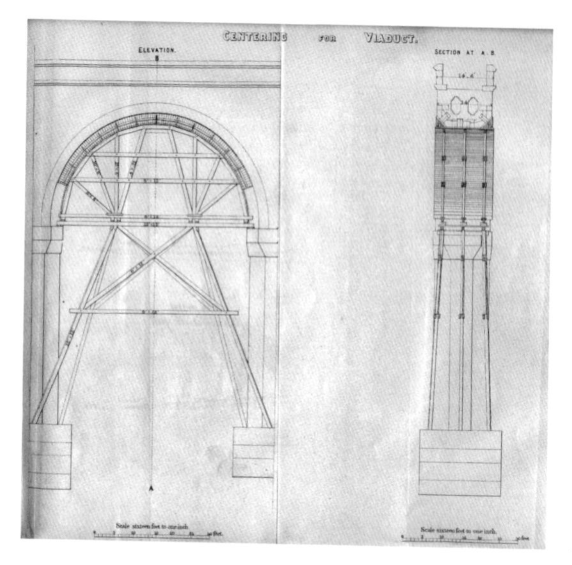

Left: *'Goed-re-Coed Viaduct, cross-section of arch and timbering'* by *Samuel Downing.*

Below: *'Goed-re-Coed Viaduct, plan'* by *Samuel Downing. (Both from Trans. Institution of Civil Engineers of Ireland, 1851)*

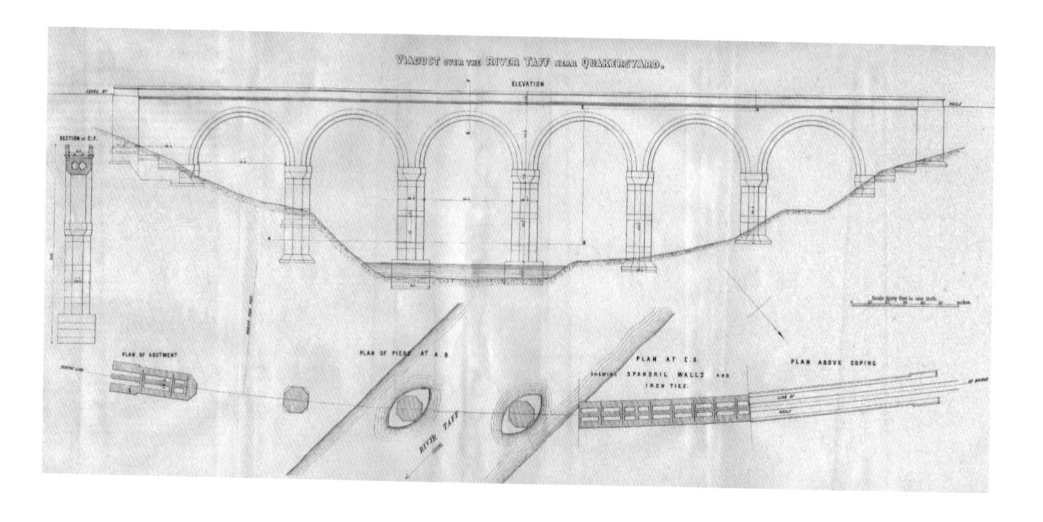

to concentrate a major part of the total rise. Even if the gradients of the line were distributed uniformly over the whole length, they would still be too severe for the operating standards of the day. The level of the railway at Merthyr was at a sufficient height to ensure adequate drainage for the station, while, at the other end in Cardiff, the railway was a few feet above the high water mark in the Bute Dock. In 25 miles the railway fell through the same height as the river – some 500ft to 520ft. It would have proved extremely difficult to keep the railway in a similar plane to the river because of the unequal nature of the upper part of the river course. Downing states that the choice was:

> '… to obtain a less elevated crossing, or dispense with the necessity of it, kept at nearly the same rate of fall as the river, the gradients would have been deemed too steep, the tunnel lengthened, and the earthwork much increased. For two miles from Merthyr, both the line and the river fall nearly at the same rate, but towards the viaduct and round this great bend in its course, the fall of the river is far too rapid to be followed by any general parallelism of the gradients, and we consequently find the line of rails considerably elevated above the stream, the average gradient thus far being about 20 feet per mile ; so that even had the line, not crossing the river, taken the circuitous direction of the valley mentioned above, it would with this gradient have had a very high and unfavourable crossing of the branch river, the Bargoed, and two other crossings over brooks, which though small streams would have given occasion to expensive works. Passing over the main stream then by this viaduct at a elevation of 105 feet, and through the short tunnel of Goed-re-coed [sic], the level of the rails is about 160 feet above the river Cynon, which must be crossed, and about half a mile distant from it, here therefore was located the inclined plane, which is worked by a stationary engine of 80 horse power surmounting 140 feet vertical in a height of 45 chains. From the foot of this plane the gradients down to the docks are of a first-class description'.[23]

Brunel had consulted the engineer, W.H. Harrison, on the merits of tunnelling or skirting and cutting through the hillside once the line had crossed the Taff. His argument for driving a tunnel 17ft-wide and 15ft-high was based on a cost of 1,100 yards at an 'absolute cost' of £4,675.[24] This was based on a cost of £9 5*s* per yard, from which stone to the value of 2*s* per ton (50 tons) could be recovered per yard, bringing the cost down to £4 5*s* per yard. This would be set against the cost of taking the line around, '… starting parallel with North entrance of Tunnel and terminating parallel with South entrance of the said Tunnel'.[25] His survey and estimate of the alternative comes to £4,714 8*s*, making the saving in favour of tunnelling just £39 8*s*, which Brunel accepted. *The Railway Times* noted that a portion of the Quakers Yard tunnel was lined with masonry, and that the dimensions of the finished tunnel were 12ft 6in in width and 16ft from the rail's top surface.[26] The shorter tunnel at Ynsicoi, some 7 miles from Cardiff, was lined with brickwork and masonry.

By crossing over to the western side of the Taff at Goitre Coed, the railway is ideally situated to tap into the Cynon valley, in particular the iron and coal works of

Looking south at Goitre Coed Viaduct, Penlock's tunnel, which is in the process of being opened out, c.1862. (Joseph Collings photograph/Stephen Rowson collection)

Looking north at the southern portal of Quakers Yard during widening, c.1862. (Joseph Collings photograph/John Minnis collection)

Aerial view of Goitre Coed Viaduct, 22 May 1963. Edwardsville is in the foreground, the deep cutting that replaced Penlock's tunnel can be seen to the left of the viaduct. (Courtesy of the University of Camridge, Air Photograph Library)

A view of the widened Goitre Coed Viaduct from upstream during the mid-1860s. The Merthyr Tramroad can be seen running under the arch on the far right. (Joseph Collings photograph/John Minnis collection)

Aberdare and Hirwain, situated on the streams flowing into the Cynon, which itself joins the Taff from the west. The main line crosses the river Cynon by a bridge of 50ft-span and two land arches of 25ft.

The line continues southwards and crosses the mouth of the Clydach valley by a 60ft elliptical arch. Downing notes in 1851 that the valley was as yet unworked. A few miles further, in the district of Newbridge, or Pontypridd as it was to become known, the Rhondda River was crossed, '... by a bold elliptical arch of 140 feet span, at a considerable angle of skew, and an elevation of 60 feet'.[27]

Newbridge was the first work to have been started on the TVR; indeed, it marked the beginning of work on the undertaking, as it was here that a foundation stone was ceremoniously laid on 16 August 1837. It was to become more than a simple ceremony as Guest had decided to publicly address the growing criticism about delays when he turned up that day with Lady Charlotte. The *Glamorgan, Monmouth and Brecon Gazette and Merthyr Guardian* reported that a vast concourse of people had turned up for the event of laying the first stone of the Rhondda Bridge, which was to mark the start of the important undertaking that was the TVR. Guest opened the proceedings and commented on the objective of laying the first stone, '... of a noble structure, which would not only be an ornament to this improving neighbourhood, but would also be the means of increasing labour, wealth, and capital, in the construction of the road.'[28] He went on to state that the directors had surmounted all the difficulties and he was glad to see the work begin, and hoped those gathered would meet again in two years' time to commemorate the opening of the TVR. Before reading out an inscription on parchment that was to be sealed in a bottle and placed under the stone, he refuted any charges of unnecessary delays on the part of the directors. The inscription read: 'The first stone of the Taff Vale Railway was laid by Lady Charlotte Guest, on Wednesday, the 16th day of August in the first year of the reign of our Sovereign Lady Queen Victoria, and in the year of our Lord 1837...'

The 'fair mason', as the *Glamorgan, Monmouth and Brecon Gazette and MerthyrGuardian* referred to Lady Charlotte, carried out the ceremony in a 'very work-man-like manner,' and addressed the meeting in a 'very neat speech', concluding with: 'I hope that this great undertaking will be the means of doing much good, not only to those who have a pecuniary interest in it, but also to those persons employed and to the districts through which it passes. I sincerely wish every success to this laudable unde taking.'

Three cheers were given for Charlotte, three for the railway and three for the engineers and contractors. Mr Waring followed as the next speaker, and the meeting adjourned to the 'Bridgewater Arms' to enjoy a cold repast. As ever, Charlotte recorded the events in her diary, which give a more amusing take on events,

> '... I went through the form of laying mortar with a pretty little trowel, but when the stone was lowered into its place, and the Engineer brought an equally interesting Liliputian hammer from his pocket for me to strike with, the idea appeared to me so absurd that I rebelled outright and insisted upon using the wooden mallet, to the no small amusement of the workmen...'[29]

The *Glamorgan, Monmouth and Brecon Gazette and Merthyr Guardian an* concluded its report by congratulating the proprietors and although this would mean that one of the most beautiful valleys in Wales would soon suffer from 'noise and dirt, smoke and steam', it felt that these feelings must '... yield to general convenience and public improvement.' A final remark about an alternative locomotive force is intriguing:

> 'It is not improbable that Giant Steam – who "now keepeth his palace" may soon yield to a Giant stronger than he and a locomotive force in a few years will be applied which will sweep away at once all that offends the ear and eye in steam conveyance.'[30]

What was it referring to? Rope haulage by fixed engines, except at inclines, had long been ruled out as a practical competitor to steam locomotion, but there was a growing interest in the development of atmospheric system of propulsion as a credible alternative. One of the early pioneers was John Vallance who had demonstrated his invention as early as 1826[31] and tried to persuade the Kensington Canal Co. to adopt his 'pneumatic railway' in 1833.[32] It was not until 1839, however, that Clegg and Samuda patented the first practical application for applying the system of atmospheric propulsion to railways. It did not turn out to be a practical proposition as Brunel was later to find out to his cost and the 'rope of air', as the Stephenson's referred to it, was confined to history. But no doubt at this time it was being talked about as a possible new 'Giant', a locomotive force, and Brunel himself may have talked about its potential. There was no reference to Brunel being at the ceremony; he was facing a growing number of commitments elsewhere.

To return to the question of the strategic crossing of the Taff, Downing argues that if the line did not cross over at the point it did i.e. Goitre Coed, and continued on the eastern side of the river, a considerable outlay would have to have been expended in carrying branches from the western mineral valleys over the river Taff to the main line. In designing the railway, Brunel did not fully take into account the business that the growing coal trade would give the TVR. Downing, albeit some years later, was confident that, without Brunel's engineering design and selection of route, '... we could not have communicated with them, and secured the traffic of these valleys, so rich in the finest description of coal'.[33] Downing does not appear to have put much emphasis on the potential of the Rhondda in this category. The original short branch into the Rhondda valley, served by a short radius curve from the TVR on the north side of the river Rhondda, was to prove inadequate and was to be replaced with a new viaduct crossing the Rhondda. To be fair to Downing, it would have been difficult to predict the great expansion in coal mining that was to take place in the Rhondda valleys, although in 1851 some entrepreneurs were taking steps to open up the celebrated No.3 seam.

In his paper, Downing is not interested in the remainder of the line, even though the line crosses the Taff three times, as these crossings were due to, '... private agreements, and special clauses, rather than to any engineering considerations...'[34]

Downing goes to great lengths to prove the soundness of crossing the river at Goitre Coed, not only at that point but also at that height and on a curve. He argued that to have continued any further without crossing would have meant the construction of a major bridge, '... nearly as high and as wide as that over the Taff at this point, on the eastern affluent, the river Bargoed, and also have recurved the line of railway northwards, in a very circuitous manner.'[35] Examination of the ordnance map of the river and its tributaries would appear to justify the crossing at this point, but not necessarily at the elevation it does. Indeed, the structure is more reminiscent of one on a cross valley and not that of a valley line – usually following one side of the river bank. A curved and high viaduct was, argues Downing, the best alternative under the circumstances, and even though he was writing only ten years after the line was opened, '... at this period of the railway system, good gradients were deemed of far higher importance than at a subsequent time.'[36]

With the decision taken to cross the river at this approximate position, Goitre Coed offered the narrowest crossing with the steepest sides. This was to be no right-angle crossing as this would have involved a deep and expensive cutting for a considerable length on the north side, while, on the south side, the tunnel would need to be considerably lengthened, and led out onto a high embankment. The resulting viaduct was to be curved in plan,

> '... with a radius of 20 chains, and a curve of the same radius, and contrary flexture, carried the line round the projecting point of land on the north side, the plan resembling the letter S; also on the south, side the curvature was continued through the tunnel, with a radius of 25 chains, the centre being on the same side as that of the curve of the bridge, thus giving the shortest length of tunnel, and a favorable position for the straight inclined plane'.[37]

It was not to be a right-angle or a skew-arch crossing but a totally Brunelian solution, his genius showing through in the design for the construction of the pier, which,

> '... completely obviates the necessity of superadding the difficulty and expense of the winding courses of a skew bridge, to those of curving in plan, which it will be seen is effected by octagonal pillars as the form of the sustaining piers, the axis of the stream being parallel to that side of the octagon which makes the same angle with the axis of the viaduct (or tangent to the curved axis rather), which the river does at this part of its course, namely 45 degrees'.[38]

Work began on the viaduct and associated works in 1837. This was to take three years and cost £25,000, including viaduct, tunnel and earthworks. Additional costs amounted to £8,000, covering the lengthening of the tunnel, undertaken because of the close proximity and position of the canal high above the railway, the extra depth required of foundations, an occupation bridge, retaining walls and the lining of part of the tunnel in masonry. The river was diverted as the foundations of each river pier was laid. As each pier was put down, the river was diverted to the opposite side of the river bed with a dam abutting on the bank and extending outwards to the line of mid-channel.

Foundations for the northernmost river pier were provided by the rock of the greyish-blue sandstone or pennant. The other river pier had to go down a few extra feet to rest on hard gravel, likewise the abutment and pier on the south valley side had to be founded deeper than expected, hence the extra costs for foundations. The other two piers and abutment on the opposite side were, like the river pier, to be solidly founded upon the rock. The rock strata that missed the southern half of the viaduct's foundations would reappear in the tunnel at an interval of around 800ft. This sandstone was to be quarried locally for works: 'Some of the beds supply excellent blocks of a good material for engineering purposes'.[39]

The particular mode of construction of the pillars or columns of the viaduct was to give a far greater appearance of lightness to the work than if carried up in rockwork. The aisler[40] or course work of the stonework allowed the band to act as one solid mass, equalizing the pressure on the pier and preventing any lateral spreading of the masonry. There was considerable detail on the construction of the column up to the bond for the arch stones on the bevilled faces. Downing states that the bond was deemed an important question, with suggestions put forward from several people, but he does not indicate whose suggestion was accepted or if Brunel determined the method that was adopted, only that: '... never at any time has there been in any part the slightest appearance of settlement, or chipping of any of the arrises...'.[41]

With the work being brought up to the springing height the centres were fixed one for each arch and it was felt unsafe to permit any arch to be keyed, and its centre transferred to another opening. The centering timberwork was put in place by laying a wall plate on the top of the cap projection of 6in on which were fixed the ends of three beams 12in-square supporting three frames of fan centering. Penry Williams includes this detail, in terms of the centering of the nearest arch, in his watercolour of Goitre Coed. The beams were then supported by dry-stone piers at around one third of their length on the slopes. The river arch beams were supported by long shores abutting on the cut waters and steadied by diagonal half-timbers springing from a pair of walings bolted on at around half their height. The illustration shows the full detail of the timberwork designed to resist the horizontal thrust inwards during the progress of the arch. Downing explains the importance of setting out the piers of curved viaducts:

> '... so as to throw all the unavoidable divergency of the radial lines upon the pier, and not any part on the arch itself, was now apparent. Three of the piers had been set out with the inner or soffit faces truly converging to the centre of the curve of 20 chains radius; the consequence was that in two of the arches the rib on the down-stream side was 4 inches greater in diameter than the uppermost, giving considerably more trouble as well in the framing as in the cutting and setting of the arch stones; those set out subsequently had the soffit faces of abutments and piers truly parallel, so that the soffit of the arch was a true cylindrical surface, and the bevilled exterior parts of a true conical figure'.[42]

There were difficulties in the construction of Goitre Coed. Downing has already commented on the problems of securing good foundations and the extra costs that had been incurred. The real difficulty in the construction, according to Downing, was the management of the spandril walls so as to gain a true uniform curvature at the string course, under the parapets at the level of the rails:

> 'For however truly the centre points of the piers may have been set in the required curve, each arch must of course lie within the vertical planes, passing through the lines strained from the outer angle of one springing quoin, to the corresponding opposite point, and thus we would have a polygon with sides of 60 feet, instead of a curve line, and we would have found it impossible to set out a line of rails suited to locomotive engines. A simple calculation show that on the concave side we had to corbel out the courses of the spandrels about 4 inches, which, small as it may seem, is more than good masons would always wish to do at such heights, and bearing in mind that the parapet projects 3 inches over the string. From the excellent quality of the stone, we were enabled to choose such binders and headers as left no room for doubt as to the stability of the structure. Trifling irregularities in carrying up different spandrels were rectified in the massive string course, in the coping finally assumed a perfectly uniform curved line. The only alterations in the original design were the substitution of a spandrel wall longitudinally through the centre line, which supported the flat covering stones, overlapping each other so as to form a base for the ballast and upper works, in lieu of the 5 feet semicircular arch, which had been proposed to effect the same object, and adding three more cross walls at each arch.'[43]

The line was completed as a single line with passing places. Of the thirteen bridges built over the line, all were constructed for a double line, but the forty-one bridges and viaducts carrying the railway over roads, rivers and canals would need to be widened when the call came to double the line.[44] It was not long in coming. In January 1846 a report entitled, 'Statement on the New Work and Contracts now in execution for the Doubling of the Taff Vale Railway' was presented to the directors. It stated that, commencing at Cardiff, the first contract was let to Hugh Allen, which is in connection with the East branch (this is the branch serving the east side of the Bute Dock). The statement goes on to say that: 'From Cardiff to Llandaff the way has been in use since the 25th of October (1845).'[45] From Llandaff, the statement refers to the current state of works such as Llandaff Canal Bridge, with mason work let to Robert Rowland and '… the girders all made…' The treatment for further bridges, including occupation bridges, embankments and retaining walls, is covered up to Ynyscoi tunnel, 'The plan decided upon by the Board to obtain a double way at the Ynyscoy [sic] tunnel is to make an open cutting. For this purpose the land is already bought.'[46] In a similar vein, the proposals for doubling the line continue to Navigation House, although, presumably on the grounds of cost, the Newbridge (Pontypridd) Viaduct is not mentioned. The Rhondda branch is noted but any further work would have to wait until after Brunel's time. The doubling of Cardiff to Navigation between 1845-48 resulted in the widening of the timber bridge at Taffs Well, the opening out of Ynyscoi tunnel and the widening of the Newbridge Viaduct. George Fisher reported on the work in progress in his report of 18 February 1847, stating that the

The rebuilt Taffs Well Bridge. Some remains of the original timber construction can be seen in the foreground. (Joseph Collings photograph/Stephen Rowson collection)

original Taffs Well Bridge, '... consisted of a wooden platform 240ft long supported in wood pile.' [47] In widening the Newbridge Viaduct no compliments would be paid to Brunel's original 110ft-skew arch. The work by Fisher was effectively a heavier structure, with a river pier supporting two smaller spans in the middle of Brunel's single span.

This left the Navigation to Merthyr as the only single line section of the TVR. With the doubling of this section and the major work of the Goitre Coed Viaduct, the view of the viaduct presented today is a different one to that presented to the artist Penry Williams. In 1862, work began on doubling the line from Navigation to Merthyr, work authorised in 1857 enabling it to '... alter, widen and improve portions of their existing Railway...'[48] John Hawkshaw was retained by the TVR to guide the Bill through Parliament and to act as consulting engineer on the construction of the works. These later works have been recorded in a magnificent set of photographs by the Cardiff photographer Joseph Collings, and go some way to addressing the lack of written and visual record of the original work.[49] With the later works, John Hawkshaw put forward his proposals regarding the opening out of the adjacent tunnel at Goitre Coed, bur not accepted by Fisher, who argued that the line of centre in the tunnel should not become the line of centre of the new track bed. Fisher did not want to move the line any closer to the line of the Glamorganshire

Canal, and wanted the tunnel wall on the canal-side to be kept as a retaining wall.[50] This was accepted with the tunnel wall buttressed to a height of approximately 20ft.[51]

With regard to the widening of the viaduct, it would have been difficult to reach a design in harmony with the original. Samuel Downing makes the following comment on the problems posed by doubling the line above Navigation, particularly with regards to the Goitre Coed Viaduct:

> 'The railroad being in the upper part a single line, nor indeed does it seem possible to extend the design, so as to be available to the increased dimensions of a double line of rails, even with the width of only 14 feet, it is evidently no easy task to turn arches upon such an impost, though by corbelling out 1 foot 2 inches, on 4 faces (Plate 1), we changed the regular octagon of the piers 5 feet 9½ inches in the side, to an irregular figure at the impost and tops of the caps of the piers, of 9 feet on the soffits and elevations, and 3 feet 6 inches on the four bevilled or coned parts of the arches : the extent to which the impost overhangs the pier is shown on the left hand pier of the plan.'[52]

Hawkshaw's design was effectively a second viaduct joined to Brunel's and, as such, compromises the graceful lines of the original. Although Hawkshaw's design is altogether thinner in terms of the piers and has wider arches within the profile of the original design, it provides a wider surface area to the river and obscures the role played by Brunel's octagonal piers. That may have been the cause of the problems of scouring that has occurred over the years, and which have resulted in the placing of stiffening girders between the two river piers at river level. Back in 1847, Fisher had also been concerned that the main arch of the Newbridge Viaduct was in danger of collapsing, and he believed his work in widening the viaduct would help to buttress and strengthen Brunel's structure. Mining subsidence has also affected the structure, with remedial work being carried out in the 1930s.[53] However, Network Rail, the body responsible for the maintenance of railway structures today, report no problems of stability with either.[54]

Gazetteer of Principal Features	Taff Vale Railway
Bute Road station, Cardiff	ST 191 748
Pentyrch crossing, Morganstown	ST 129 818
Ynyscoi tunnel (opened out)	ST 118 847
Newbridge Viaduct, Pontypridd	ST 070 901
Llancaich Incline Viaduct	ST 080 938
Llancaich Branch (incline top)	ST 084 943
Main line Incline	ST 085 951 - 090 960
Goitre Coed Tunnel, Quakers Yard (opened out)	ST 090 962
Goitre Coed Viaduct, Quakers Yard	ST 087 963
Dowlais Railway Junction	SO 053 052
Site of Plymouth Street station	SO 051 055

Chapter 8 Notes

1 Proceedings of the Institution of Civil Engineers, Vol.1, printed in full in Vol 5.
2 Pugsley, Sir Alfred (ed.), (1976). p.95. *The Works of Isambard Kingdom Brunel An Enineering Appreciation* Chapter V, Owen, J.B.B. London: Institution of Civil Engineers/University of Bristol. Paper presented by Thomas Macdougall Smith.
3 MacDermot, E T, revised by Clinker, C R. (1964). p.49. *History of the Great Western Railway*, Vol.1. London: Ian Allen. The other engineer; Nicholas Wood, agreed with Brunel's remedy.
4 Pugsley, Sir Alfred. ed. (1976). Owen, J.B.B. p.95.
5 Pugsley, Sir Alfred. ed. (1976). Owen, J.B.B. p.100.
6 University of Bristol, Special Collections, IKB Letter Book. 2, p.79, n.d.
7 The *Glamorgan, Monmouth and Brecon Gazette and Merthyr Guardian*, 11th March 1837. This was the full title of the publication at this time but one that was to change over the years, for brevity in the text it will be referred to as the Merthyr Guardian.
8 University of Bristol, Special Collections, IKB Letter Book 1. p. 250. Recommendation of William Lewis, 11 September 1837 and notice to commence contract No.3, p.253. 18 September 1837.
9 *Glamorgan, Monmouth and Brecon Gazette and Merthyr Guardian*, 8 April 1837. University of Bristol, Special Collections, IKB Letter Book 1. p.248. Notice to commence contract No.2, 3 August 1837.
10 *Glamorgan, Monmouth and Brecon Gazette and Merthyr Guardian*, 2nd September 1837.
11 *Some South Wales Collieries (Illustrated), No. 10, The Great Western Colliery Company Ltd*, p.88-89, in 'The Welsh Coal Fields', a series of articles originally published in *The Syren and Shipping* and repinted in 1906, The Syren and Shipping Ltd: London.
12 Downing, Samuel. (1851). pp.23-48. *Description of the Curved Viaduct at Goed-re-Coed near Quakers Yard, Taff Vale Railway*. Trans. Institution of Civil Engineers of Ireland. Vol 4, Part 1.
13 Jones, Jack. (1951). River out of Eden. p.13-16. London: Hamish Hamilton Ltd.
14 Jones, Jack. (1951). p.37-38.
15 Barnes, M. (2000). *Civil Engineering*. 138 August 2000. pp.135-144. Paper 12109.
16 Pugsley, Sir Alfred. ed. (1976). Owen, J.B.B. p.91.
17 Gibbs, George Henry. ed by Simmons, Jack. (1971). [p.67]. *The Birth of the Great Western Railway, Extracts from the diary and correspondence of George Henry Gibbs*. Bath: Adams & Dart, Bath.
18 Downing, Samuel. (1851). p.31.
19 Downing, Samuel. (1851). p.31.
20 Pugsley, Sir Alfred. ed. (1976). Owen, J.B.B. p.92.
21 Downing, Samuel. (1851). p.25.
22 Downing, Samuel. (1851). p.25.
23 Downing, Samuel. (1851). p.27.
24 University of Bristol, Special Collections, TVR Facts Book p.121. Arguments for tunnel, W.H. Harrison. I am indebted to Steve Rowson who believes that this W.H. Harrison is William Henry Harrison, a land and mineral surveyor of Merthyr, and not the clerk and surveyor of the Glamorganshire Canal, William Harrison (*d*.1848). As both the tunnels on the TVR combined amounted to 377 yards in length, Harrison's measurement must have been square yards. W.H. Harrison was to later survey a number of mineral lines for the Marquess of Bute.

25 University of Bristol, Special Collections, TVR Facts Book p. 121. Arguments for tunnel, W H Harrison.

26 *The Railway Times*, pp. 137-38. This is based on the second or unpublished volume of 'The Railways of Great Britain and Ireland' by Francis Whishaw Esq., Civil Engineer, M. Inst. C.E.

27 Downing, Samuel. (1851). p.26.

28 *Glamorgan, Monmouth and Brecon Gazette and Merthyr Guardian*, 19 August 1837.

29 Bessborough, Earl of. ed. (1950).p.56.

30 *Glamorgan, Monmouth and Brecon Gazette and Merthyr Guardian*, 19 August 1837

31 Snell, J. B. (1971 reprinted 1973). p.116. *Railways: Mechanical Engineering*, Longman Group Ltd 1971, Arrow edition 1973, London.

32 Ottley, George. (1965 second edition 1983). Ref. 296. *A Bibliography of British Railway History*, Science Museum/National Railway Museum, London.

33 Downing, Samuel. (1851). p.26.

34 Downing, Samuel. (1851). p.26.

35 Downing, Samuel. (1851). p.25.

36 Downing, Samuel. (1851). p.26.

37 Downing, Samuel. (1851). p.28.

38 Downing, Samuel. (1851). p.28.

39 Downing, Samuel. (1851). p.29. Downing refers to the Report of the Geological Survey, Vol.1. p.192.

40 For aisler read ashlar; hewn or squared stone used for building purposes.

41 Downing, Samuel. (1851). p.29.

42 Downing, Samuel. (1851). p.30.

43 Downing, Samuel. (1851). p.31-32.

44 *The Railway Times*, p.138.

45 PRO. RAIL 684 56. 5 January 1846.

46 PRO. RAIL 684 56. 5 January 1846.

47 PRO RAIL, TVR Report by George Fisher, 18 February 1847. Rowson, Stephen. (1999), p.11.

48 Jeffreys Jones, T.I. ed. (1966). p.98, *Acts of Parliament Concerning Wales 1714-1901*. Cardiff: University of Wales Press.

49 See the article in the *Archive* magazine, Issue 21, March 1999, by Rowson, Stephen; *Engineering on the Taff Vale Railway 1861-8*. The Lightmoor Press: Lydney.

50 PRO RAIL, TVR Report by George Fisher, 12 February 1861.

51 Rowson, Stephen. (1999), p.12-13.

52 Downing, Samuel. (1851). p.28. Plate 1 is from this and is reproduced on p.144, 'Goed-re-Coed viaduct plan'.0

53 Henderson, Frazer. (1991), p.30. *The Railway Engines and Architects of Wales: Aberystwyth*. Fig.4 on p.30 shows supporting timbers (echoing the original timber centring) in place in 1931.

54 Correspondence with Alan Evans of Network Rail, August 2004. He points out that this does not mean that problems were not encountered at the time of construction or widening, and that there was some concern about mining subsidence in the area of Quakers Yard in the 1970 and 1980s [the strengthening girders at river level being considerably older than this].

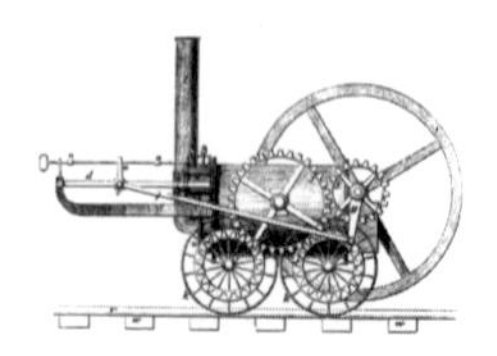

9

Openings And Blockades

'... Get the Railway Finished as Soon as May Be After My Dock is Open' [1]

Lord Bute (John Crichton Stuart, second Marquis of Bute, 1793-1848) supported the construction of the Taff Vale Railway insofar as it would provide trade for the dock he was proposing to build in Cardiff. This, the first enclosed dock to be built at Cardiff, was completed in 1839 and, at the time of its construction, was the largest masonry dock in the world. It covered 19½ acres and cost Lord Bute £350,000, of which £222,757 was in cash and the remainder the value of the timber, stone, brick and other materials from his estates.[2] He expected to recoup this investment through the resulting increase in value of his estate and the exploitation of minerals on his land that would be shipped through the Bute Dock.

Naturally, Bute sought to secure the maximum advantage for his investment and was prepared to use every power at his disposal to achieve this, a position that was to cause great friction between him and the TVR, who did not want to commit itself to shipping solely at the Bute Dock. If they were forced to do so by financial constraints or opposition, the TVR felt it should receive the most favourable terms as compensation for making it their sole outlet. The TVR directors, under Guest's leadership, had at an early stage decided to keep their options on shipping outlets as wide as possible. Part of this strategy, put forward by Walter Coffin in 1835, was to have three outlets to the sea, namely at the mouths of the Taff, Ely and Rhymney Rivers. The Taff outlet would be the main terminus of the TVR, linked to the Bute Dock, which they argued should be under their operational control with all the wharfs leased to the railway. Bute interpreted this as an attempt by the promoters, '... to make it a matter of favour or threat whether they will bring their railway to my port or not.'[3] These comments were made to his agent, E.P. Richards, on 9 November 1835, and a month later he was of the opinion that the Taff Vale Railway company was made up of little more than a '... combination of Merthyr ironmasters and tradespeople. We should find a company of strangers much more reasonable to deal with.'[4]

In the serious game of negotiation, some concessions were to be made by the TVR promoters; the notice for the TVR Bill in November 1835 referred to a branch railway, '... with proper wharfs, staithes, works and conveniences...' to Cogan Pill.[5]

The Bill, however, that made its way through Parliament in 1836, contained no references to shipping places on the Ely or Rhymney, and Bute saw no reason to oppose it, even investing in it. Shares were taken out by Bute in the name of E.P. Richards, John Bird and Lord James Stuart (brother of Lord Bute), in an attempt to create his own pressure group within the predominant rump of Bristol shareholders. Guest and Bute faced each other at opposite ends of provincial politics, and would be unlikely partners, but each saw the merits of each other's undertaking. There was a long history of antagonism between Bute and Guest but Bute recognised the importance of the TVR being completed without delay; '... it cannot be denied that it is an advantage to me to get the railway finished as soon as may be after my dock is open.'[6] There was to be some conflict, however, in terms of the labour required to build the respective works. The novelist Jack Jones relates the story of the building of the Bute Dock in *River out of Eden*, and includes a fictional discussion with Lord Bute and the dock contractor, Daniel Storm, over the strike by navvies employed on the building of the dock, in which Storm exclaims, 'We're already paying them more by a ha'penny an hour than the contractors making that railway up in the hills are paying their navvies.'[7] Storm was referring to the predominantly Yorkshire labourers who had staged what was the first workers' strike in Cardiff, which was broken by bringing over 200 Irish workers to work on the building of the dock.

Bute Road station, Cardiff docks, in 1986. The building was then used as a railway gallery by the Welsh Industrial and Maritime Museum and the headquarters of the Butetown Historic Railway Society. (SKJ photograph)

In the year before this episode, Guest had made a public speech at the laying of the foundation stone for the Newbridge Viaduct.[8] There, on 16 August 1837, Guest had invited those gathered to meet again in two years' time to commemorate the opening of the TVR, but this would not be possible. The report of the TVR's half-yearly meeting on 23 August 1839 reported that contracts 2, 3, 4, 6, 7, 8, 10, 15 and 16 were finished with, apart from one or two road bridges and a small quantity of earthworks. The directors regretted that there had been so much delay with regard to '... the main line between Crockherbtown and the terminus near the Bute Ship Canal, unexpected difficulties arose in obtaining possession of land...'[9] This was clearly down to intransigence and delaying tactics on the part of Lord Bute, the principal landowner in Cardiff. The cartographer John Woods, was to make the comment, 'Most of the Town & Environs belong to the Marquess of Bute.'[10] His map, published in the late 1830s, shows the TVR line crossing over Bute Street to terminate a short distance away from the Glass Houses, owned by Guest. This may have been wishful thinking on the TVR's behalf as Bute was concerned about keeping the line as close to his dock as possible. The Glass Houses were one of the few places in Cardiff not under Bute control.

To be fair to the Bute camp, land acquisition was not the only cause of delay. Work on Contract No.1, the Goitre Coed Viaduct, was not finished and the work, which was to take three years, had not started until 1837. Additional work, which included extra depth for the foundations, lengthening of the tunnel and retaining walls, accounted for almost a third of the total cost. It was to result in the Taff Vale Railway enjoying two openings of its main line: a partial opening from Cardiff to Navigation (Abercynon) on 8 October 1840, and the opening throughout upon the completion of the Goitre Coed Viaduct on the 21 April 1841.

It was originally intended to have a private opening of the first section of the line on 7 October 1840, with the railway opening to the public on the following day.[11] This was decided at a director's meeting held on 25 September 1840, with Sir John Guest in the chair. Tickets for the private opening would be sent to all the shareholders. The trains would run at 8 a.m. and 4 p.m. from Cardiff, with the trains returning at 9 a.m. and 5 p.m. respectively, from Navigation House. It was to be left to the chairman, engineer and secretary to make the necessary arrangements.

However, another meeting had to be held the following day, at which Sir John was not present. Walter Coffin took the chair, and the secretary reported:

> '... that unfortunately an old engagement precluded Sir John and Lady Charlotte Guest being present at the intended opening of the Railway on the 8th October and the Directors feeling the imperative necessity of the Public Opening being not later than the 8th which is the earliest day the Line can be opened to the Navigation House'.[12]

It was resolved that as the whole of the main line was likely to completed in around three months, the celebration intended for the 8th would be deferred until then, '... when the Directors hope Sir John and Lady Charlotte Guest will honor [sic] them with their presence'.

Sir Josiah John Guest, 1785-1852. (SKJ collection)

Lady Charlotte Guest, 1812-1895. (SKJ collection)

1 Hacqueville parish church and the monument to Marc Brunel. (SKJ photograph)

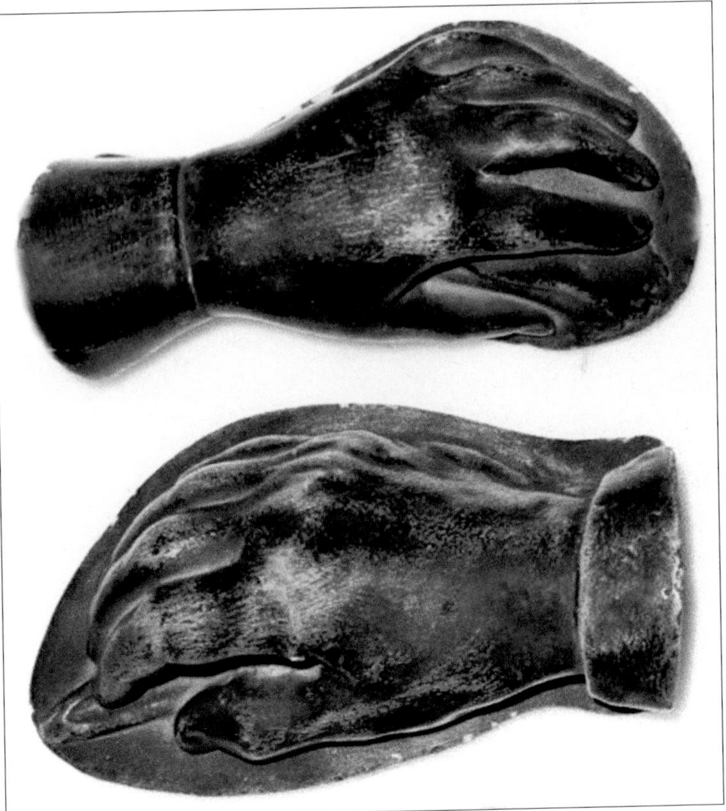

2 Plaster cast of Marc and Sophia Brunel's hands (the casts were originally donated to Cardiff Museum in the nineteenth century). (Courtesy of the National Museum and Galleries of Wales)

3 Two recent views inside the Thames Tunnel. One shows the tunnel after the rails were lifted for refurbishment, and the other, the restored and relined tunnel in 1997. (Courtesy of the Dr G. Sauer Co. Ltd)

4 Grosvenor Bridge, Chester. (SKJ photograph)

5 Statue of Isambard Kingdom Brunel on the Victoria Embankment in London. (SKJ photograph)

6 'The Vale of Taff' by Thomas Horner, 1819. This part section of a panoramic watercolour by Horner (1785-1844) shows the canal running through rural countryside below present-day Pontypridd. The figure leaning on the canal bridge is believed to be the Duke of Sutherland (Horner's patron). In 1803 Malkin commented on the scene, '... The road from Newbridge (Pontypridd) to the Duke's Arms, a respectable inn, passes along the riverside, with which a very curious canal keeps pace. This canal is esteemed a remarkable instance of art triumphing over the obstacles thrown in its way by nature'. (Courtsey of T.E. Morgan, OBE, DL)

7 Interior of the Rolling Mill at Cyfarthfa works by Penry Williams. Note the outline of Cyfarthfa Castle on the right. (Courtesy of Cyfarthfa Castle Museum)

8 Neath Abbey, showing horse drawn trams and a horse-worked 'gin.' See map on p.51. (From E. Donovan, *Descriptive Excursion through South Wales and Monmouthshire*, 1804)

9 Merthyr Tramroad below Pontygwaith. An interesting section of the tramroad today on the section between Mount Pleasant and Quakers Yard. The route of the tramroad is clearly defined with stone sleeper blocks still in situ along much of the way. The holes and indentation of tram plates on the sleeper blocks can be clearly seen. Note also the profile of the track bed cut into the hillside. (SKJ photograph)

10 Pontygwaith Bridge over the tramroad, half-a-mile south of Mount Pleasant. The bridge crossing the tramroad here carried a very steep road down the side of the valley to the site of an ancient iron furnace. The name 'Pontygwaith' comes from this association, meaning 'Works Bridge', and refers to the original bridge crossing the river Taff.
(SKJ photograph)

11 Upper Tramroad Bridge, Quakers Yard. Travelling downstream, this was the first of two bridges that carried the Merthyr Tramroad at Quaker's Yard. Originally of wooden construction, they were both replaced by elliptical stone arches following the collapse of the upper bridge in 1815. (SKJ photograph)

12 The Penydarren locomotive, a full-size working replica of Trevithicks's locomotive at the Welsh Industrial and Maritime Museum (WI&MM), Cardiff, where it was built and inaugurally steamed on 10 June, 1981. Work on building the replica was initiated and led by Dr Stuart Owen-Jones of WI&MM. (SKJ photograph)

13 'The Dawn of a New Era: No.1' by David Weston. The painting shows Trevithick's Penydarren locomotive, the man himself, his Cornish mining background and the historic letter written to Davies Giddy (Gilbert). (Courtesy of Sir W.H. McAlpine and David Weston)

14 Navigation Hotel, Abercynon. A few yards from the site of the Trevithick monument were the Glamorganshire Canal's offices, now the Navigation public house. Following the closure of the Abercynon colliery, the road layout shown in this late 1970s photograph has been changed. (SKJ collection)

15 New Trevithick mural at Beacon's Place, Merthyr. The new artwork was officially opened on 21 February 2004 at the 'Tesco' store development in Merthyr. (SKJ photograph)

16 Re-dedication of the Abercynon momument, 21 February 2004. This monument, standing in front of Abercynon fire station, marks the end of Trevithick's historic run to the canal basin. (SKJ photograph)

17 Union Bridge across the Tweed near Berwick-upon-Tweed. The bridge links England and Scotland with Welsh iron, and was opened in 1820. The bridge was designed by Capt. Sir Samuel Brown RN with chain work supplied by Brown's Newbridge chainworks. (SKJ photograph/ Chainbridge collection)

18 Menai Suspension Bridge. Crossing the Menai Straits to Anglesey, it was part of Thomas Telford's road improvements from London to Holyhead for Irish traffic. (SKJ photograph/Chainbridge collection)

19 *Left:* £2 Trevithick coin issued in 2004 (SKJ photograph)
20 *Right:* Taff Vale Railway uniform button. The design may be based on the Sharp 'singles' supplied in 1840 (see p.181) and the locomotive on high-level branch line (on p.171). (SKJ photograph)

21 The original Egyptian-style design for the cast-iron pier cladding on the Clifton bridge, 1831 watercolour by Samuel Jackson. (Courtesy of the City of Bristol Museum and Art Gallery)

22 Suspension chain links on the Clifton bridge. Two thirds of the chain links had been originally supplied by the Dowlais Ironworks (as bar-iron) for Brunel's Hungerford Suspension Bridge. (SKJ photograph/Chainbridge collection)

23 An 1860 lithograph of the old Cardiff Town Hall in the High Street with a statue to the Second Marquess of Bute in front. (SKJ collection)

24 A modern (2004) view of the facade of the former 'Angel Inn' in Cardiff, from the walls of Cardiff Castle. This was the venue for the first General Meeting of the proprieters, following the Act of Incorporation of the TVR in 1836 (see p.108). The Wales Millennium Stadium can be seen in the background. (SKJ photograph)

25 Alltwen Tramroad, Aberystwyth. The tramroad, presumably built by George Bush, supplied stone for the building of Aberystwyth harbour. (Courtesy of the National Library of Wales)

26 Newbridge Viaduct, Pontypridd, 1986, Brunel's 110ft skew arch over the Rhondda River. (SKJ photograph)

27 'Goitre Coed Viaduct', 1841. The talented local artist, Penry Williams, was supported by the Merthyr ironmasters, even to the extent of him receiving financial help from both the Crawshays and the Guests when he went to Rome to continue his art studies in 1827. Apart from occasional visits to his native Wales, Penry Williams was to stay and work in Rome until his death. Presumably it was on one of these visits that he made his preliminary sketches of Goitre Coed Viaduct, which was completed back at his studio in Rome. His watercolour shows the slender octagonal columns of Brunel's viaduct with the contractor's timber centering still in place in the nearest arch. The Merthyr Tramroad passes under that arch and a locomotive can be seen on it. Despite Trevithick's pioneering run of 1804, locomotives were not reintroduced until 1832, when the Dowlais Iron Co. ran its locomotive *Perseverance* to the canal basin at Abercynon. (Courtesy of the Elton Collection of the Ironbridge Gorge Museums).

28 'Goitre Coed Viaduct', 1845. Another watercolour by Penry Williams shows the whole length of the viaduct from river level. The viaduct comprised six arches, each of 50ft-span, built upon octagonal columns of masonry 14ft in diameter. The maximum height above the river level was 100ft. There is a slight rise in the level of the viaduct, with the train climbing towards Merthyr. The watercolour is inscribed: 'Penry Williams, Rome, 1845'. (Courtesy of the Elton Collection of the Ironbridge Gorge Museum).

29 A modern view of the Goitre Coed Viaduct, looking downstream. (SKJ photograph)

30 The 1840 date stone on the Goitre Coed Viaduct. (SKJ photograph)

31 The main line above Abercynon. The line was effectively split into two sections for locomotive haulage with a rope-hauled incline necessitating engines being stationed at both ends of the line. Apart from the inconvenience and delays this was to cause, it was not seen as posing any serious problem as the bulk of the traffic would be travelling downhill towards Cardiff i.e. descending in the direction of the load. The incline embankment is on the right with modern traffic using the 1867 diversion, which was built to permit easier gradients for working by the more powerful engines of that period. (SKJ photograph)

32 A bridge for farm access under the old incline. This farm-access bridge was removed to the extension of the dual carriageway trunk road (A470) to Merthyr. It is interesting to note that this underbridge was widened in 1862 as part of the doubling of the line, yet this section was abandoned when the 1867 diversion was built. Note the slight difference in arch profile between Brunel's 'half' and the 1862 work; Brunel's work is characterised by deep chamfering of the arch. In the background is the access bridge for the diversionary line which comes back to the original route just before the Goitre Coed cutting (originally a tunnel) a few hundred yards on. (SKJ photograph)

33 The Iron Duke broad-gauge replica at Cathays Yard, Maindy, Cardiff, part of the NRM tour of 1985. Although designed by Brunel, the TVR was never a broad-gauge railway. A length of broad gauge was laid to accommodate the replica at what had always been a standard-gauge railway works. (SKJ photograph)

It is interesting to speculate as to how Sir John had forgotten about this 'old engagement' the day before; was it possibly because it was one of Lady Charlotte's? There is no reference to any other engagement in the published edition of Lady Charlotte's journal, but immediately preceding the TVR advertisement, in the *Glamorgan, Monmouth and Brecon Gazette and Merthyr Guardian* on 3 October 1840, announcing the opening of the railway to Navigation on 8 October, there is an advertisement announcing the publication of *The Mabinogion Part 3*, by Lady Charlotte Guest on 7 October.

The Cambrian announced: 'It is appreciated that the opening as far as Newbridge will take place next week, two of the carriages, built by Walter Williams of Bristol having arrived on Wednesday last and six more are expected within a week. The first class carriage cost 400L each.'[13] As it turned out, the railway was opened on 8 October and it was an event that was recorded by Lady Charlotte in her journal as both she and her husband were present. It turned out to be a busy day, as the following extract shows:

> 'October 8… Off in the open carriage before eight to join the Railway train. Again the weather was perfectly fine and the whole day went off very well and proved very enjoyable. We waited some time at the Navigation House for the engine to arrive. It had been detained by getting off the Rails in a part of the road only lately laid. In the meantime I enjoyed all the preparations and all the bustle which was going on, and which presented certainly rather a curious scene, the arrangements being anything but complete… When the train arrived we took our places in it and proceeded to Cardiff. We had a slight detention on the road owing to the train again getting off the Rails, and we had various stoppages in various parts of the line to take up passengers. We arrived however before twelve, at which hour the great opening train set off and we proceeded again by it back to the place whence we had started. This time the train went forward without accident or delay, and on reaching the [station] we at once set forth to walk up to the Goitrecoed bridge and tunnel ; Merthyr and I headed the party at first up the Incline plane, but when we came to the tunnel we made the band of music, which had accompanied us, go first.'[14]

On getting to the tunnel, Lady Charlotte described the effect produced by Bush who had caused the tunnel, '... to be lighted up by numerous candles stuck in the sides of the rock, and the effect was exceedingly striking and picturesque.' Even more exhilarating was emerging out of the tunnel and onto the viaduct;

> 'The Viaduct looked as beautiful as ever as we issued forth from the tunnel upon it. We should have lingered longer to examine it but the time was running fast away. So they gave us God Save the Queen and gave cheers for the Railway &c &c., among which was one for me, and then walked back to where the train was waiting. We had a delightful walk, which some, however, had considered rather a severe one. Anything equal to the brilliancy of the valley I never saw.' [15].

The tunnel was to be opened out to accommodate the laying of an additional line of track in 1863, although one side of the tunnel wall remains to this day. In terms of the logistics for the opening, further changes had been made to the directors' resolution of 26 September 1840, since at that meeting Bush had pointed out that he felt that there was some danger with the train returning as late as 5 p.m. from Navigation House. It was resolved to start the afternoon train an hour earlier so that the train could return at 4 p.m. Bush was obviously concerned that there could be problems, which are borne out by Lady Charlotte's comments in the extract above about the train coming to pick them up, '... getting off the Rails', and a further derailment she experienced on the train returning to Cardiff.

When Lady Charlotte's train returned to Cardiff, she noted that there was, '... an immense party' waiting to make another trip up the line. The Guests' decision to attend on the 8th appears to have led to the decision to combine the running of a special train on the same day as the public opening. A fuller description of the opening was reported in *The Cambrian* on 17 October 1840:

> 'Opening of the Taff Vale Railway
>
> The public opening of the completed portion of this interesting and valuable line between Cardiff and Navigation House took place on Thursday se-night, and was celebrated by the inhabitants of Cardiff, Newbridge, and other places in its neighbourhood in a manner which showed that they were fully alive to the local and commercial importance of the undertaking.
>
> It had been arranged by the Directors, in order to avoid confusion and to prevent interference with the ordinary traffic, as well as to afford shareholders an opportunity of deliberately inspecting their line, that the special train should leave Cardiff at 12 o'clock, and accordingly, long before that hour, the immediate neighbourhood of the station, and, indeed, every point from which a view of the proceedings could be obtained, was densely crowded with spectators.
>
> Never did Cardiff present a more gay and animated appearance. All classes of the inhabitants seemed to vie with each other in demonstrations of joy and satisfaction. The bells rang merry peals, flags streamed form the castle steeples, and the houses of such of the inhabitants as were located in the neighbourhood of the station. Cannon were fired continuously throughout the days, and the very excellent band of the Glamorganshire Militia, which was stationed at the terminus, contributed to the holiday by performing a variety of marches, quadrilles, etc.
>
> At a few minutes before 12 o'clock eight carriages were appended to "The Taff" locomotive engine, and shortly afterwards the company, to the number of 180, took their seats. All the rank, fashion and beauty of the neighbourhood were there.
>
> The shrill tones of the steam whistle soon gave notice to the spectators that the carriages were about to start, and in a minute afterwards the whole train moved slowly forward and majestically for some yards, greeted by loud cheers, salutes of artillery, and accompanied by the lively peals from St. John's belfry, which latter commenced in the morning with the rising sun, and continued at intervals till night.

Goitre Coed cutting, replacing the tunnel which, on the opening day in 1840, was '... lighted up by numerous candles.' (SKJ photograph)

Shortly after the start some portion of the distance was performed at the rate of forty miles per hour. The speed was then necessarily decreased, so as to cause the arrival of the train at Newbridge, a distance of 12 miles, precisely in 31 minutes. At Newbridge a small portion of the passengers alighted, but by far the greatest part proceeded as far as the extremity of the railway, near the Navigation House, where the numerous band of F. Crawshay, Esq., was drawn up, and received the approaching train with "God save the Queen" and "Rule Britannia". Descending from the carriages, the passengers, preceded by Sir John Guest, entered a temporary wooden house erected by the proprietors, who very spiritedly supplied the party with champagne and other wines and viands in abundance.

After the lapse of a few minutes, it was announced by an individual who ascended the table, that all persons who wished to proceed to the tunnel at "Cefn Glas", would be allowed half-an-hour to walk there consequently a large party set forth, preceded by Sir John Guest, Lady Charlotte and the Crawshay band.

Upon their arrival at the tunnel, everybody was amazed at the glittering novelty, for it had been most tastefully illuminated to the whole extent, upwards of 300 yards. After remaining there and being highly gratified for some time, and enjoying an advantageous view of the six arched bridge beyond, the party returned, and as soon as all were seated in the carriages, they advanced slowly to Newbridge, whence they arrived in Cardiff in 28 minutes, the weather having been most propitious during the whole day.'[16]

Ariel view of Plymouth Street station, Merthyr. The former TVR station is to the right of centre; the former VNR running across the top of the photograph to the High Street station. (Courtesy of Merthyr Tydfil Reference Library)

The Cambrian also reported that: '... the Chairman expressed his regret at the unavoidable absence of Mr. Brunel, and concluded by proposing the health of the resident engineer Mr. Bush. Three times three. Mr. Bush in acknowledging the compliment said he felt deeply indebted to the shareholders and the directors.' The *Glamorgan, Monmouth and Brecon Gazette and Merthyr Guardian* [17] also reported on the opening, with the reporter taking the opportunity to comment on the recent decision by the Parliamentary committee that examined the 1840 TVR Bill, to disprove the Ely branch; '... the advantages of the Ely branch were again attempted but ineffectually'. The clauses relating to the 1840 Bill were dropped but the preamble on the desirability of a dock on the Ely was passed. There was also a public dinner in honour of the opening, starting at 4 p.m. in the 'Cardiff Arms', to which anyone interested could attend on the purchase of a 6s-ticket from the bar of the 'Cardiff Arms'.

Brunel was busy elsewhere; the GWR was being opened in stages and in the period up to July 1840, sections from Twyford to Faringdon Road had been opened. In its 29 August 1840 edition, *The Cambrian* announced that the, '... GWR to be opened Bath to Bristol on 31st inst... one of the engines went at a speed of 60 miles an hour, or a mile a minute.' The TVR had achieved a maximum speed of 40mph on its opening day, with the special train covering the 12 miles between Cardiff and Pontypridd in 31 minutes. Fortunately, the 12mph speed restriction in its original Act had been repealed by the Taff Vale Amendment Act of 23 July 1840.

An alternative to the official celebrations taking place in 1840 was a bare-knuckle fight between Shoni Sgubor Fawr (Johnnie Great Barn) of Penderyn, and John Nash,

Plymouth Street station, Merthyr, in use as a National Carriers depot. The former VNR line and viaduct can be seen in the background. (Courtesy of Merthyr Tydfil Reference Library)

Site of Plymouth Street station, Merthyr, 1978. This was the original terminus of the TVR in Merthyr, comprising '... the usual convenience of passenger shed, carriage and goods' sheds and warehouse. There are also workshops, engine houses for four engines, coke sheds, weighing machines, etc etc'. In the middle foreground there is a double engine house, with a stable block to the right. The Vale of Neath Railway Viaduct is on the left. The TVR routed its passenger services over the viaduct to share the VNR High Street station from 1878. Plymouth Street continued to be used as a goods station until 1967, and the whole site has been redeveloped twice since 1978. (SKJ photograph)

Engine sheds at Plymouth Street, Merthyr. (SKJ photograph)

the champion of Cyfarthfa and a railman in the works. Shoni was a famed pugilist and one time 'Emperor' of Merthyr's notorious red-light district known as 'China'. He had once been on the side of law and order as a soldier in the 98th Regiment of Foot. He had infiltrated and gathered information on that extreme group of industrial agitators known as the 'Scotch Cattle' of Monmouthshire. Ironically, while the pugilistic event he took part in was part of the celebrations for a new form of transport, he was to play a leading role in disrupting an older system, the turnpikes, during the Rebecca riots in 1843, for which he was eventually arrested.[18]

It was to be more than three months later that the line was to be opened throughout, notice being given that the entire main line was to be opened on 21 April 1841. On the Monday before, George Bush, together with Joseph Ball, accompanied Sir Frederic Smith on a trial run over the unopened section between Navigation and Merthyr. They stopped for a short time at the 'Angel Inn' in Merthyr before returning to Cardiff. As the Inspector General of Railways, Sir Frederic reported his decision by letter;

> 'Cardiff 20 April 1841
> Sir – I have to acquaint you for the information of the Directors of the Taff Vale Railway Company, that I shall this day report to the Lords of the Committee of Privy Council for Trade that I am not aware of any objection to the Opening of the line to Merthyr Tydfil to-morrow, provided the arrangements for working the Inclined plane shall be then complete; and I shall have great satisfaction in making known to their

Lordships the very satisfactory state of the line and in bearing testimony to the skilful manner in which Mr. Bush has grappled with the professional difficulties he has had to contend with in the formation of this railway.
I am, Sir, your obedient servant
Frederic Smith
L. Col. Royal Engineers
Inspector General of Railways
To J Ball'

The letter is reproduced by the *Glamorgan, Monmouth and Brecon Gazette and Merthyr Guardian*, which, in its coverage of the opening throughout of the main line, picks up the comment that it was, '... highly creditable to the talented engineer Mr. Bush.'[19] Guest is present but there is no reference to Brunel.

A further notice in the same newspaper for 17 April 1841 gave the following timetable and announced that: 'The entire main line will be opened to the public on Wed 21 April.'

From Cardiff		From Merthyr	
8.00	Morning	8.20	Morning
11.00	Morning	11.35	Morning
3.00	Afternoon	3.20	Afternoon

A subsequent notice on 1 May 1841 gave the arrival times and the re-designation of the 11.00 up and the 11.35 down trains as mixed trains conveying passengers and goods, with the cancellation of the 8.20 down train from Merthyr.[20]

Up From Cardiff

8.00	Morning Arrives at Merthyr Tydfil 9.25
11.00 Morning Mixed	Arrives at Merthyr Tydfil 12.45
3.00	Afternoon Arrives at Merthyr Tydfil 4.25

Down From Merthyr

11.35	Morning Mixed	Arrives at Cardiff 1.20
3.20	Afternoon	Arrives at Cardiff 4.25

In 1841 the first excursion into the Rhondda branch was opened in June from the main line at Pontypridd to Dinas, or Aerw (Eirw) as the TVR referred to it. It was 4 miles and thirty-eight chains long and was not to penetrate any further until 1855. Also in 1841, the Llancaiach branch, 3 miles and twenty-nine chains long, was opened from Stormstown Junction on the main line, into the Llancaiach district.

Of the northern branches, only the Dowlais branch was to be completed in line with the original Taff Vale Railway proposals. Guest wanted a connection from his works to the TVR at the earliest convenience, and had ensured that this was in the 1836 Bill, but Parliament only allowed for a branch to '... communicate with the Tramroad leading to the several Ironworks of Pen-y-darran, Dowlais and Plymouth...'[21] This was rectified in 1837, but the powers were allowed to lapse. Fresh powers to construct this branch were renewed by the TVR in 1840 but these also lapsed and had to be renewed in 1843. As with the 1840 Act, the Dowlais Co. had the right to take possession of the works and finish the line if it was not completed by the TVR. This actually happened in 1848 when the TVR started work on the branch, but made very little progress, and the Dowlais Iron Co. were forced to take charge. It was apparent that it was impossible for them to complete the line in the timescale permitted under the powers of the 1843 Act. The Merthyr to Cardiff turnpike still had to be crossed and the line was not even marked out.[22] Dowlais was forced to apply for its own Act of Parliament to build a line from the TVR to its works.[23] Why Dowlais had allowed things to get to this stage, being let down by the TVR on several occasions, is difficult to understand. Guest had played a pivotal role in the building of the TVR and was its first chairman (although the issue of the Dowlais branch had led him to resign from the position of chairman in 1840, to be re-elected following the conclusion of a new agreement). There was also the memory of Dowlais losing out on a direct communication with the Glamorganshire Canal over half-a-century ago, when the branch canal originally proposed was dropped. Such a branch would have stretched the frontiers of canal engineering to its limits. To keep it supplied with water alone, the canal engineer, Dadford, had calculated that in 1¾ miles the lockage involved would be no less than 411ft.[24] This led to Dowlais joining forces with Penydarren to build a railroad, shortly after the canal received its Act of Parliament in 1790, although the line, built under the 4-mile clause' as the Dowlais Railroad, was only to serve the interests of Dowlais.[25]

The uncertainty over the renewal of the principal lease of the Dowlais works and the real possibility that Guest would have to leave Dowlais in the 1840s was an obvious distraction. In April 1843 it appeared that Lord Bute would not go through with the agreement he had made through Stephenson and, by March 1847, Guest felt he had no option but to dispose of his plant.[26] Guest resigned from the chairmanship of the TVR for good in 1847, a decision that drove Lady Charlotte to cry all afternoon, '... I love that Railway for old associations' sake. It was his creation, and through almost overpowering difficulties he carried it through, and made it prosper.'[27] Quite unexpectedly, and mainly due to the death of Bute in 1848, a settlement was reached on 21 April 1848, just ten days before the expiry of the lease. In the following year Guest obtained his own powers, under the Dowlais Railway Act of 1849, to construct a railway from the uncompleted works started by the TVR, to terminate inside the works itself.

In the year of the Great Exhibition, which a great number of Dowlais workers were to visit, the Dowlais Railway was finally opened. On 21 August 1851, the railway was opened; the first section of the line, in which it climbed 330ft over

66 chains of its 1 mile 72 chains length, was operated by incline.[28] Two R & W Hawthorn twin-cylinder steam engines drove the incline cylinder at the top of the incline, with the remainder of the line operated by locomotives. Guest was a sick man by this time but insisted on travelling from London to participate in the opening of the branch. Because of his ill health, Guest was forced to leave early and Lady Charlotte took over the proceedings.[29]

The engineer commissioned by Bute to construct his dock at Cardiff was William, later Sir William, Cubitt (1785-1861). The dock was to be initially to be known as the Bute Dock or Ship Canal and later, following the construction of further docks in Cardiff, as the Bute West Dock. Brunel had met Cubitt at Cardiff in October 1834[30] but there was little common ground between Cubitt and Brunel, although later Cubitt would be a supporter of the atmospheric system of propulsion which he was to adopt on the Croydon Railway. As a civil engineer, Cubitt had considerable experience of canal, river improvements and dock works, and from 1842 he was engaged on deepening and improving the navigation of the upper reaches of the river Severn. However, Brunel does not appear to have regarded Cubitt highly, and in February 1838 he wrote to Anthony Hill regarding a meeting of the TVR directors in Cardiff, saying he had also written to Cubitt, 'But [because of] the very cautious manner with which any thing is treated by the Marquis and his Agents, I do not anticipate any result.'. The problem seems to have been how to work out the precise nature of the link between the TVR and the dock. In March 1838, Brunel wrote impatiently to Cubitt:

> '... I have unfortunately been unable to see you and am now off for Wales and ... shall hope to see you [in Cardiff] and that the appointments may have been made by others - it is a question which seems to me must be of importance to both parties, the more so that we must decide much.'

In October 1836, Brunel wrote about the proposed Ely docks, '... I have now considered the question of the practicability of constructing Docks at the proposed Termination of the Ely Branch...' He goes on to state that he has got a general survey of the river and had undertaken some borings:

> 'From the general opinions entertained by those who possess experience from all the information I have been able to obtain from my own observations and from the results of the surveys the soundings and the examination which has been made there appears to be no doubt that as regards the requisites for a good harbour - depth of water - sheltered position - facility of ingress and egress - and anchorage, the mouth of the River Ely is superior to that of the Taff.
>
> These natural advantages are not newly discovered but from all that I can learn it seems to have been a generally received opinion notwithstanding the present position of the Town and the Sea Dock - notwithstanding the choice of situation for the proposed new Docks; that the River Ely ought to have been originally selected for the site of the harbour and still offers great inducements.

Such is the received opinion and I believe a well founded one of the natural advantages of the River and the position. The extent and the nature of the ground at the terminus of this branch of the railway fully enable the Company to avail themselves of these advantages.

In the space between Cogan Pill and Penarth Point by enclosing a portion of the Bay formed by a sweep of the River a spacious Dock may be formed sufficient for all the trade which has been calculated upon for the proposed railway and capable of being extended should any considerable increase of trade hereafter require it -…' [31]

Brunel refers to a plan, a copy of which is not in the letter book, and goes on to describe the docks as having a tidal basin and inner dock, with the entrance pointing upwards (upstream), closed by a single pair of gates. An additional entrance may also be formed by a lock and a double pair of gates from the river. At high tide both entrances might be in use. The size of the dock is given at 14 acres, of which '… the south side of the docks alone would give about 20 berths…' With expansion, 80 berths could be accommodated, and the inner dock could contain up to 200 vessels. The tidal basin would cover 2 acres and would be capable of holding fifteen vessels averaging 200 tons each. To keep the dock clear of mud and silt, Brunel proposed to get rid of such deposits, '… by a simple process which has been adopted with perfect success in the Bristol float…' Brunel estimated the cost of such a dock at £60,000: '… the amount is trifling compared with the certain advantages secured to the Trade of the Railway by the addition of such an accommodation to the terminus.' There was, however, a legal problem confronting the company, as neither the 1836 Act nor any later Acts allowed for the Ely dock plan. In 1840 the TVR had been forced into dropping the clauses from its Bill, but the preamble referring to the desirability of the Ely docks was left in and passed by Parliament. During the Bill's passage through Parliament, it was argued that, 'The Ely would be an open and free port while the new harbour was the private property of the marquess of Bute.'[32] Despite such arguments, Bute and his agents conducted a successful propaganda campaign against the Ely dock and, by August 1840, Robert Stephenson, acting as an advisor to Bute, believed that the Ely dock was dead.

The arrangement between the railway line and the Bute Dock was not a workable one until 1842. Bute had been in contact with Robert Stephenson since 1832, employing him in the early 1840s to advise and report on engineering matters, Cardiff being the first dock project he was associated with. Brunel had frequent meetings over the dock issue and the location of the railway terminus with Stephenson. In February 1840, he wrote to Robert Stephenson on the subject of the latter, and enquired if Monday at 12 o'clock was convenient, '… to you at your office or mine.'[33] A few days later he was writing to Bush to inform him that he and Robert Stephenson would be in Cardiff on Saturday 22 February 1840, and would devote, '… that day and the next to the Cardiff Docks & Ely – collect any information you can upon both – particularly the former – of course you will meet us.[34] Stephenson also reported on the state of the Bute Dock following the failure of part of the dock wall through faulty workmanship in the spring of 1840. The main contractor, Daniel

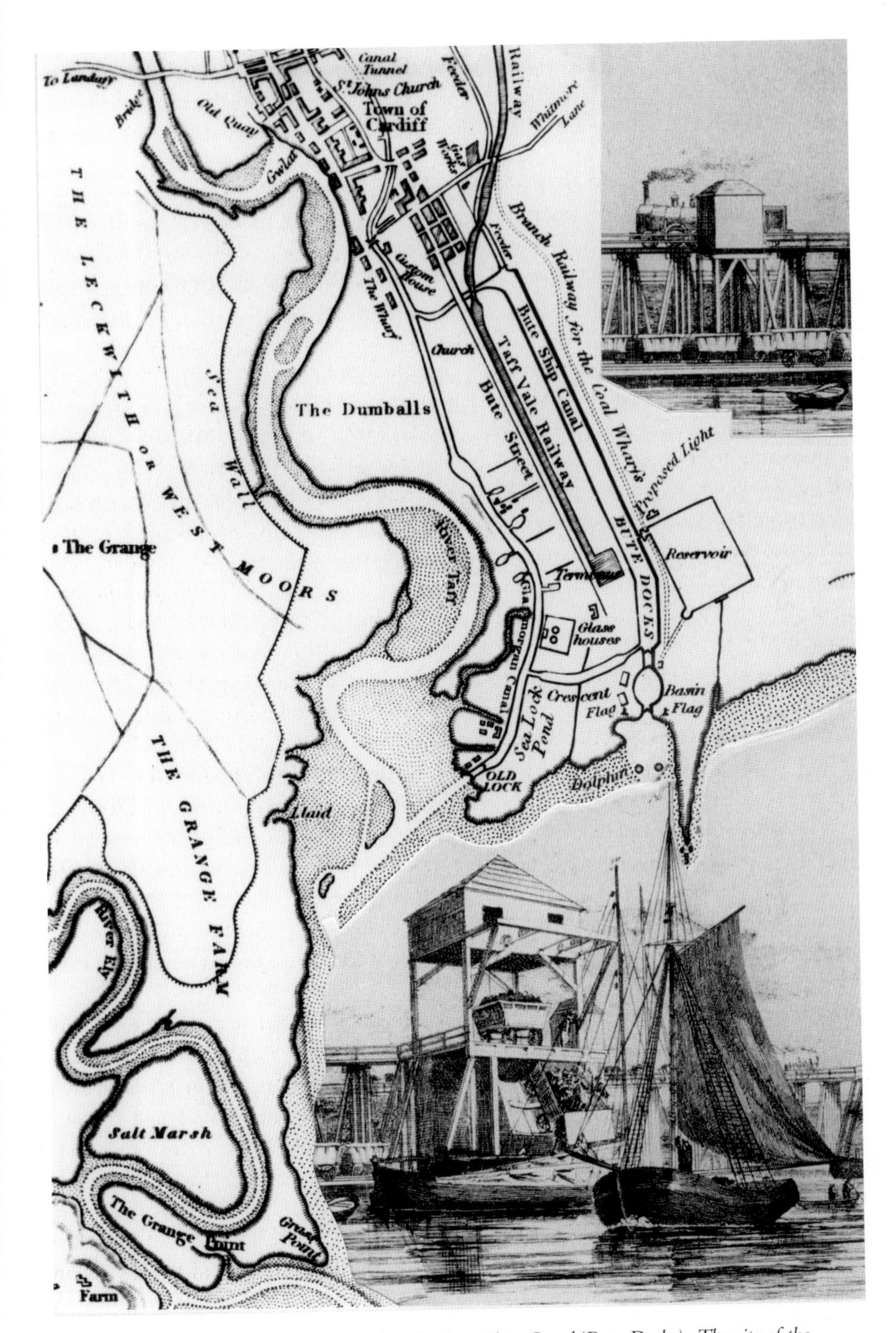

The TVR in Cardiff and the Bute West Dock (Bute Ship Canal/Bute Docks). The site of the proposed Ely Harbour can also be seen (bottom left-hand corner) along with the high-level coaling appliances (overlaid on the map). (From Capt. W.H. Smyth's Nautical Observations, *1840)*

Storm, was to blame. He was also engaged on TVR contracts and his bankruptcy was to cause as many problems for the TVR as it was for Lord Bute.

Professional relationships with Lord Bute were not easy, and he suspected disloyalty in everyone, resulting in a consequent large turnover of staff and consultants. Cubitt was forced out 'for alleged extravagance' and Capt. W.H. Smyth was dismissed in September 1841.[35] Capt. Smyth, who had been placed in overall control of the Bute Dock operations in 1839, published his Nautical Observations on the Port and Maritime Vicinity of Cardiff in the autumn of 1840. Published at Lord Bute's instigation, it appeared at face value to be a description of the Bute Dock, but actually set out to denounce the Ely project as an attempt by the TVR directors to force access to the Bute Dock on their terms. Another example of the Bute propaganda against the Ely dock even extended to Lord Bute persuading Stephenson to publish arguments against the scheme.[36] Despite this show of loyalty, Bute was to lose confidence in Stephenson and the connection ceased in the mid-1840s. The position with Brunel and Stephenson at Cardiff was an early professional encounter, although they had previously found themselves on opposite sides with Bute in the middle. Lord Bute had engaged Stephenson in connection with the Stanhope & Tyne Railway and sought his advice on his own interests in the Brandling Junction Railway. When that railway constructed a branch to Monkwearmouth in 1839, Brunel was also involved.[37]

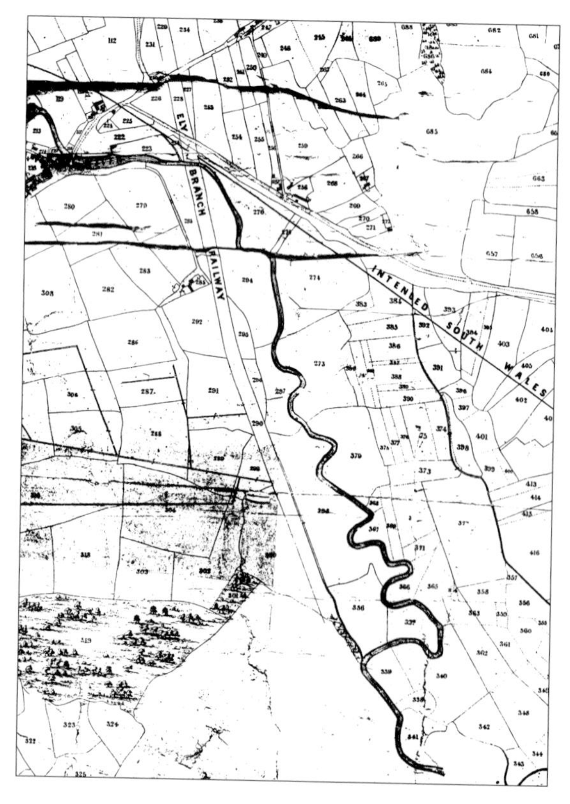

Tithe map of 1846 showing the original Ely branch, for which land had been bought, marked out and fenced. The line runs through the Ely hamlet (Ely Bridge) across the line of the South Wales Railway towards the mouth of the river Ely. (Courtesy of Ely Old & New collection)

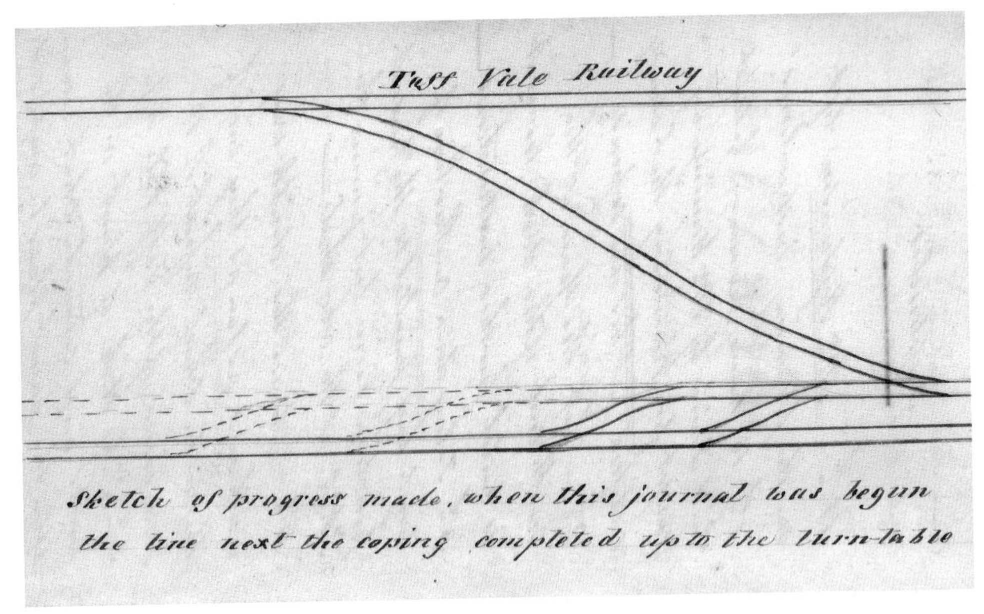

TVR sidings alongside Bute Dock, from a plan in Henry Swinburne's diary dated 18 September 1841, Ref. ZSW 539/4. (Courtesy of Northumberland Record Office and Sir John Browne-Swinburne)

Stephenson appointed Henry Swinburne (1821-55) as resident engineer to oversee the work to provide rail access to the west side of the Bute Dock. As a consequence of the limited access allowed by Bute in this period, the TVR was to make increasing use of the 'Little Dock' on the Glamorganshire Canal. The Little Dock had been constructed on land owned by Bute for which the TVR had powers to lease under the 1836 Act. In 1840, Bush was instructed to look at facilities for shipping coal and iron from the Little Dock.[38] With regular iron and coal traffic beginning in June 1841, the TVR began to tranship from the rail-side to both the Bute Dock and the Little Dock, at first by manual methods, namely wheelbarrow and shovel. Things improved with the signing of an agreement in March 1842.[39] Early in December 1842, the TVR adopted one of Stephenson's recommendations and installed the first of a number of wagon lifting and tipping appliances on the south-west corner of the Bute Dock.[40] This hydraulic appliance was to be later transferred to the Little Dock. S.W. Allen recalled that the Little Dock had several tips on the south side, the wagons being brought in on the level of the street which they crossed, and he described the operation of the tips as follows:

> '... sunk into the ground in the centre of each tip, was a very large cylinder and ram, on the top of which was a cradle, on which the laden trams rested. Pipes of large diameter communicated with these rams and the bottom part of the big elevated tank, so that the simple weight of the water, due to its height and the large diameter of the rams, was quite sufficient to provide the necessary power to lift the laden trams the required height in order to allow the coal to slide out into the vessel's hold.'[41]

Brunel does not appear to have designed any shipping appliances for the TVR but later designs in his sketchbook appear to have been used by the South Wales Railway at Cardiff, and Bullo Pill on the Severn.[42] Because the access to the Little Dock was across Bute land, it was the subject of much aggravation, particularly in the summer of 1843. In that year, the first of several discussions and provisional agreements took place between Bute and the TVR over the purchase or lease of the Bute Dock. Against this background, the TVR decided to revive the Ely project, putting it on the agenda at its fifteenth half-yearly meeting, held in August 1843. An agreement was reached in November 1843 for the lease or purchase of the Bute Dock, but this was not welcomed by all, particularly William Crawshay, who saw a union between the TVR and Bute as a threat to the future of the Glamorganshire Canal. To counter this, he suggested the merger of all three concerns – railway, dock and canal – but Crawshay's demands eventually led to collapse of these proposals and, in the meantime, the TVR shareholders had rejected the agreements made by the directors.[43] The possibility of such a merger had got as far as including the suggestion in a Bill for the 1844 session of Parliament, but in December 1843 this part of the TVR Bill had to be deleted as Bute withdrew from the agreement, only to later change his mind. Consequently, negotiations with Bute were to drag on and, in November 1844, a special meeting of the TVR was called, and held at the 'White Lion Inn' in Bristol, to consider a number of proposals. Bristol had become the regular venue for TVR meetings soon after the opening of the line because of the dominant control of the Bristol-based shareholders. Guest was unable to attend the meeting due to the illness of his children and, when Lord James Stuart (brother of Lord Bute) and Robert Roy (Bute's business advisor) arrived, there were outbursts of 'The Ely, The Ely!' from the shareholders. After discussion, an outcome was reached, with Mr W.D. Bushell, a director of the TVR, reading out a proposed heads of agreement:

> 'The Ely Branch to be abandoned upon terms agreed upon. The money paid for the land on the Ely branch to Lord Bute to be returned to the Company and the land required on the east side of the Bute Docks for the railway to be granted to the Railway Co., for an equivalent to be paid annually as rent to be settled by Mr. Stephenson and the railway made by them to Mr. Stephenson's satisfaction.'[44]

It was also reported that Mr Stephenson said that he thought Lord Bute ought not to charge for any part of the land on which the line may be made opposite the frontage of the of the dock, but the land from the north of the dock to the junction should be paid for.[45]

It was carried unanimously by the shareholders, and in May 1846 an agreement was signed which finally put paid to the TVR constructing a dock or erecting staithes on the river Ely, or indeed for shipping from any dock other than Bute Dock. On 26 August 1846, this agreement between Bute and the TVR became law, being incorporated in an Act of Parliament by the TVR to avoid any further changes of mind by Bute. The move to the east side of the dock turned out to be an advantage to the TVR as it allowed for an improved type of hydraulic coal-tipping appliance to be employed. This appliance had to be elevated above the level of the dock side, and the TVR built a high-level branch line to the east side of the dock (the East branch). The first of these tips were successfully tested on 22 July 1848, when a train of coal from the Cymmer colliery was tipped into the vessel Mayflower, the branch being officially opened on 4 December 1848.[45]

Although a working arrangement had finally been reached and ratified by an Act of Parliament, it did not mark the end of disputes between the TVR and Bute. Brunel's selection of the mouth of the river Ely as a suitable shipping place was to be borne out by the later development of Penarth dock, and also one that had a connection with Brunel through the work of his son, Henry Marc Brunel, who was to progress to a larger project in the shape of the Barry docks.

Chapter 9 Notes

1 Davies, John. (1981). p.281. *Cardiff and the Marquesses of Bute*, Cardiff: University of Wales Press.
2 Glamorgan and the Bute Estate, John Davies PhD thesis, U C Swansea, 1969, p.575, quoted from Captain Smyth, p.21.
3 Glamorgan and the Bute Estate, John Davies PhD thesis, U C Swansea, 1969, p.595.
4 Davies, John. (1981). p.279.
5 Davies, John. (1981). p.279. I am indebted to Colin Chapman who informs me that the deposited plans for the Bill which show a proposed 'cut' at Cogan Pill.
6 Davies, John. (1981). p.281.
7 Jones, Jack. (1951). *River out of Eden*. p.13-16. London: Hamish Hamilton Ltd.
8 Pontypridd was adopted as the name for Newbridge in the 1860s.
9 *Glamorgan, Monmouth and Brecon Gazette and Merthyr Guardian*, 23 August 1839.
10 Wood, John. (late 1830s). *Map of Cardiff*. Reproduced by the Glamorgan Record Office.
11 Directors Minutes 25 September 1840.
12 Directors Minutes 26 September 1840.
13 *The Cambrian* 26 September 1840.
14 p.116-17, Lady Charlotte Guest Extracts from her Journal 1833-1852, edited by The Earl of Bessborough, John Murray, London, 1950.
15 p.117, Lady Charlotte Guest Extracts from her Journal 1833-1852, edited by The Earl of Bessborough, John Murray, London, 1950.
16 *The Cambrian*. 17 October 1840. Opening of TVR.
17 *Glamorgan, Monmouth and Brecon Gazette and Merthyr Guardian*, 10 October 1840.
18 Taff Vale Railway. Extract from: 'The Rebecca Riots', David Williams, University Of Wales Press, Cardiff, 1955. p.247. See also; Western Mail 2 Feb 1942. BN 619.
19 Published in The Glamorgan, Monmouth and Brecon Gazette and Merthyr Guardian, 24 April 1841.
20 *Glamorgan, Monmouth and Brecon Gazette, and Merthyr Guardian*, 1 May 1841.
21 Taff Vale Railway Act, 6 Wm. IV, cap. Ixxxii, 21 June 1836.
22 Rattenbury, Gordon & Lewis, M.J.T., (2004), *Merthyr Tydfil Tramroads and their Locomotives*, p.42-3, Railway and Canal Historical Society, Oxford.
23 Act lxi (L & P), 12-13 Victoria, 1849.
24 Rowson, Stephen & Wright, Ian L., (2001), *The Glamorganshire and Aberdare Canals*, p.28, Black Dwarf Publications, Lydney.
25 Rattenbury, Gordon & Lewis, M.J.T., (2004), p.11-12.
26 Jones, Edgar, (1987), *A History of GKN, Volume 1*, p.111-13, Macmillan Press, Basingstoke.
27 Jones, Edgar, (1987), p.103 and Bessborough, Earl of, (1959), *Lady Charlotte Guest, extracts from her Journal 1833-1852*, p.186, John Murray, London.
28 Rattenbury, Gordon & Lewis, M. J. T., (2004), p.43.
29 Guest, Revel & John, Angela V., p.128 , (1989), *Lady Charlotte A Biography of the Nineteenth Century*, Weidenfeld and Nicolson, London.
30 Brunel archives, Brunel University, Brunel office diary DM1758, 23 October 1834.
31 Brunel archives, Brunel University, Letter Book 2, pp.75-78, Taff Vale Railway, Ely docks, 24 October 1836.
32 Davies, John. (1981). p.280-81, *Cardiff and the Marquesses of Bute*,University of Wales Press, Cardiff.

33 Brunel archives, Brunel University, Letter book 2b, p.41. 7 February 1840, Brunel to R. Stephenson.
34 Brunel archives, Brunel University, Letter book 2b, p.41. 11 February 1840, Brunel to George Bush.
35 Davies, John, Glamorgan and the Bute Estate, Chapter VII, part 1, Communications and the Bute Estate, p.581-82, Ph. D Thesis, University College, Swansea, 1969.
36 Davies, John. (1981). p.280-81.
37 The Butes had been involved in the development of the northern coalfields for many years. In 1796 the Most Noble John Marquis of Bute was one of the parties that had entered into an agreement with Boulton & Watt for a winding engine at Pontop Pike Colliery. See I*ndustrial Revolution: a Documentary History. Series One: The Boulton & Watt Archive and the Matthew Boulton Papers from Birmingham Central Library Part 5: Engineering Drawings - Crank, Canal, dock & Harbour, Mint, Blowing, Pumping and other engines,* c.*1775-1800.* Adam Matthew Publications Ltd, Marlborough, http://www.ampltd.co.uk /collect/p169.htm The proposal by the GWR to share Euston as its London terminus also brought the two engineers together.
38 Powell, Terry, *The Taff Vale Railway at Cardiff Docks*, in the *Welsh Railways Archive*, Vol II, No. 6, November 1997. Rowson, Stephen & Wright, Ian L., (2004), *The Glamorganshire and Aberdare Canals, Volume 2.* Lydney: Black Dwarf Publication. See Chapter 11; *The Taff Vale Railway Company's Little Dock*, pp.269-78.
39 Mountford, Eric R (1987). p.10. *The Cardiff Railway*, Oxford: The Oakwood Press.
40 Rowson, Stephen & Wright, Ian L., (2004), p.271.
41 Allen, S.W. (1918), p.43. *Reminiscences*, Western Mail Ltd, Cardiff. The dock was filled in or around 1916, Allen talks about it being 'in the last year or so'. Mountford, Eric .R (1987). p.12, refers to three of these old type coal lifts being re-erected at the Little Dock.
42 Brunel archives, Brunel University.
43 Davies, John. (1981). p.281-82.
44 *Glamorgan, Monmouth and Brecon Gazette, and Merthyr Guardian*, 16 November 1844.
45 *Glamorgan, Monmouth and Brecon Gazette, and Merthyr Guardian*, 16 November 1844.
46 Mountford, Eric R (1987). p.12.

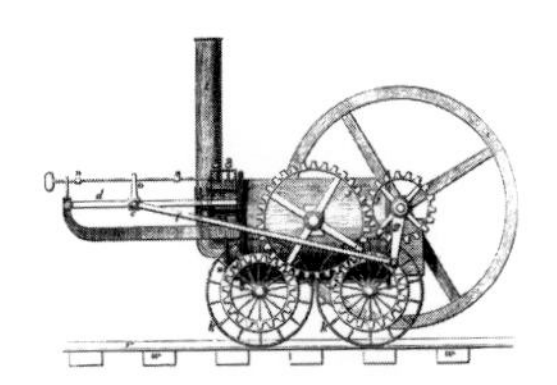

10
Working The Line
'There You Are, At It Again, Mr. Brunel, A-Breaking the Rules'[1]

Breaking the rules came naturally to Brunel, particularly those rules drawn up by the railway companies for their employees. As an example, in 1849 he asked his resident engineer, Francis Fox, then engaged on the construction of the South Wales Railway, to write to George Fisher as he had caused a TVR employee to break one of its many rules and regulations:

> 'Mr. Brunel requested me to inform you that he took the liberty of riding on one of your Engines, yesterday, from Merthyr and being contrary to your rules, and the Engine Driver being anxious that you should be informed of it, Mr. B wished me to tell you that no blame attached to the man, and that he will feel obliged by your not noticing this apparent neglect of duty.'[2]

The earliest existing book of rules and regulations is dated 'February 1853'[3], although rule books would have been issued to employees shortly after the opening of the line in an attempt to cover every contingency possible in working the line and operating the timetable. The rule broken would have been the equivalent of the 1853 rule, No.380: 'Not any person, except the proper Engineman and Fireman, is allowed to ride on the Engine, or Tender, without the special permission of the General Superintendent.' It is not known if the general superintendent, George Fisher, was to accede to the request by 'Mr. B' to ignore this breech of the rules. In 1846 Fisher halted the progress of a new TVR 0-6-0 engine (Newbridge) being tested on the main line by the locomotive superintendent, Alexander Colville, on the grounds that the bridges were not safe, and further attempted to suspend the driver with him![4]

No less than 2,120 rules were included in the 1853 edition, which the company expected every employee to memorise and apply where necessary; the fine for not keeping a copy of these regulations, together with train tables, on the employee's person while on duty was 5*s* (rule No.27). The rules also served to remind the employee that the TVR ran its trains on the right, contrary to the convention adopted by nearly every other railway company; 'It must be borne in mind, by every person in the employ of the company, that the proper road is always the right hand

The original Cardiff station at Crockherbtown, later Queen Street (building with smoke vent). It was taken in 1886 prior to its demolition. (SKJ collection)

road, looking in the direction he may be travelling.' (No.59) The roles and correct procedure of every employee involved in the running of the line were covered, as were the various system of signals used. Employees were forbidden by rules and regulations to interfere with, '... matters of a political nature' (No.241), and even to annoy one another under pain of a fine (No.263). The TVR was concerned with the spiritual wellbeing of its staff, requesting that every employee, '... on Sundays and Holy Days, when he is not required on duty, that he will attend a place of worship; as it will be the means of promotion when vacancies occur.' (No.26) The TVR even went to the extent of issuing bibles, kept in station waiting rooms, but clearly expected its workers to think more deeply about the duty they owed to the company in the last of its general regulations (No.284): 'In the morning think what thou hast to do; and at night ask thy self what thou has done'.

Passenger traffic on the TVR had begun with the opening of the line on 8 October 1840, from Cardiff to Navigation. With the prolonged negotiations with Bute, mineral traffic was slow to start, but the opening of the TVR's two mineral branches in June 1841 helped to accelerate this. From this time, a regular traffic of iron, iron ore and coal began on the railway and a working timesheet was published. However, the real impact of the potential of coal traffic was not to be felt in terms of traffic carried and revenue earned until the problem of access to the Bute Dock was resolved.

It had also been expected that the Glamorganshire Canal would lose a considerable amount of its traffic with the opening of the railway, but this loss was not as dramatic as feared, because sufficient business was available for both concerns. Indeed, the canal was to carry more iron and iron ore than the railway for over a decade after the railway was opened. This is all the more surprising as the *raison d'être* of the TVR had been the transportation of iron. By contrast, the amount of coal carried on the railway was, from 1848, roughly twice as much as that carried by the canal. This date coincided with the resolving of the dock arrangement and a rapid increase in the coal trade, which provided more tonnage for the canal and a growing customer for the TVR.

Years.	Conveyance.	Iron. Tons.	Iron Ore. Tons.	Coal. Tons.
1846	G.C.	145,781	42,531	287,271
	T.V.R.	61,067	....	....
1847	G.C.	156,693	55,081	262,077
	T.V.R.	64,256	....	....
1848	G.C.	155,015	56,924	281,967
	T.V.R.	60,026	55,000	500,000
1849	G.C.	156,196	69,745	245,785
	T.V.R.	70,671	60,000	510,000
1850	G.C.	167,379	82,680	268,631
	T.V.R.	67,603	60,000	560,000
1851	G.C.	190,633	96,408	294,537
	T.V.R.	74,701	51,000	580,000
1852	G.C.	217,319	80,525	301,829
	T.V.R.	77,176	60,000	650,000

Iron, iron ore and coal sent by the Glamorganshire Canal and Taff Vale Railway, 1846-1852. (From Jelinger Symons 'The Industrial Capacities of South Wales', in the Cambrian Journal, *Vol.1, 1854.*

South Wales was to become the greatest steam-coal-exporting centre in the world and the Rhondda valleys, the single most important area of production. Exploitation of the Rhondda began with Dr Richard Griffiths in 1790, and continued with Walter Coffin. Coffin established the reputation of the Lower Rhondda as the finest bituminous coal in South Wales. In 1839 he sent down 56,000 tons of coal, a figure that represented over a quarter of the total coal tonnage carried by the canal in that year.

Other speculators moved in to exploit the bituminous seams, one of the most prominent being John Calvert, the man who had reluctantly moved to Wales as a contractor on the TVR. It was not until 1851 that the deeper steam-coal measures began to be exploited, heralding the Rhondda's dominance in steam-coal production. This impetus was responsible for the extension of the Rhondda branch, into the Rhondda Fach, and from Dinas up to Treherbert by 1856.

The question of the gauge had been a major decision; by adopting the narrow or standard gauge, Brunel appears to have chosen, or allowed, a different type of rail to be used in preference to the 'bridge' type favoured for his broad-gauge lines. *The Railway Times* noted that the rails used were of parallel form, weighing 55lbs to the yard, and fixed in chairs by compressed wooden keys of pitch pine timber.[6] Costing around £8 10*s* per ton, they were supplied by three ironworks, the Pentwyn Iron Co. of Pontypool and the Dowlais and Plymouth works of Merthyr Tydfil. The fixing of the chairs to the sleepers was of 'a novel description' and enabled the them to be fixed by screw bolts prior to the sleepers being sent to form the line, ensuring the accuracy of the gauge. This would appear to have been an innovation of Bush's, as his view on this mode was that it offered three advantages; it was a more secure fastening than spiking the chair; the gauge of way could be accurately established; 'Thirdly that it obliges the platelayers to straighten or curve every rail as the case may be, as they have no opportunity of twisting the rails by the spiking of the chairs'.[7]

The TVR operated as a single line when it was opened and trains were worked on a time-interval basis, giving plenty of scope for problems. There are many examples of locomotives jumping the rails or simply breaking down, with the result that the engine of the next train would be pressed into use to get the broken down train out of the way. There were only six passing places on the line. Passenger trains were hauled by the Sharp 'singles'; these were locomotives built by Sharp Roberts & Co. of Manchester. They had 5ft 6in-driving wheels and were capable of pulling trains of 25 to 35 tons. Carriages varied greatly between classes and, from the beginning, the doors of passenger carriages were kept locked when the train left the station and were only opened if the passengers requested that they wanted to get out. The carriage builder, Williams of Bristol, had supplied the first batch of second-class coaches in September 1840. George Bush was prompted to write to the makers on 2 October 1840: 'I request that you will send over locks for all our second class carriages and some person to fix them, as our second class passengers will be troublesome customers if we do not lock them up'.

It is not known if Bristol train travellers were deemed to be less troublesome than their Welsh counterparts, but it was assumed that Bristol craftsmen were better at their trade. Mr W. Davies recalled the standard of some of the early work carried out on the TVR by reference to a panel once owned by TVR carpenter, Philip Lucas:

Thwaites and Carbutt 0-6-0 locomotive. Built in Bradford, three of the engines were sold at Cardiff in 1866 and may have been used by the contractors engaged on widening works (see page 146). (From Alfred Rosling Bennett, The Chronicles of Boulton's Siding, *London, 1927.*

Ely*, a 2-4-0 locomotive built by Kitson & Co. of Leeds. Kitsons supplied* Ely *(No.33) and* Rhymney *(No.34) to the TVR in 1854. From* The Railway Magazine*, July 1898, interview with Ammon Beasley, general manager of the TVR.*

> 'It was a panel with the T.V.R. arms beautifully painted, representing the red dragon with the motto "Y ddraig goch a ddyry gychwyn" overhead, and "Cymru fu a Chymru fydd" underneath. These panels were designed for decorating the T.V.R. passenger carriages. When the carriages were made, a poor local Welshman was engaged by a Bristol firm to assist in their decoration, and in particular to paint this design. The Bristol workmen would not have him to work on the same side of the carriages as themselves, which was considered the front; but put him to paint the back of the carriages. The late Mr. Fisher, when he came to examine the work, found the Welshman's work was far superior to the Bristolians; so that the Welshman's side had to be the front, and the Englishmen's the back.' [8]

In September 1841 *The Cambrian* announced that:

> 'It is appreciated that the opening as far as Newbridge will take place next week, two of the carriages, built by Walter Williams, of Bristol having arrived on Wednesday last and six more are expected within a week. The first class carriage cost 400L each.'[9]

Working the line on a time-interval basis meant that an interval of either 5 or 10 minutes would be enforced between trains. The TVR employed its own police force to maintain time-interval working and to operate the disc and crossbar signals at stations and junctions, the disc and crossbar signal being very much a Brunelian feature. Time-interval working operated for some thirteen years until the electric telegraph was introduced in 1854. It is interesting that the TVR did not adopt the telegraph earlier, as one of the early telegraph pioneers was Edward Highton, an engineer appointed as general superintendent to the TVR when George Fisher left.[10] Highton had taken out a joint patent with his brother, the Revd Henry Highton, in 1844, which was tried out on a 10-mile stretch of the London & North Western Railway (L&NWR) around this time.

In a book written in 1852, Highton explains that, 'The acceptance by the author of an appointment as General Superintendent to a railway company, and his being engaged in other engineering works, prevented the further carrying out of this plan of telegraph'.[11] Highton went on to take up the post of telegraphic engineer on the L&NWR, being replaced on the TVR by George Fisher on 23 December 1845.[12] As an MP, Guest had sat in the various Parliamentary debates over the electric telegraph.

The TVR promoted the private use of its telegraph, as a handbill dated 9 July 1860, advertises that the electric telegraph could be used for private messages. Messages could be sent by the company's wires to and from ten principal stations. They would be charged 9d for up to twenty words and then 3*d* for each additional ten words or fraction thereof. Semaphore signals were introduced soon after the introduction of the telegraph and some were in use at the Cardiff end of the line as early as 1855.

The TVR's engine shed was close to the Little Dock that was used to bring in their early locomotives. Bush managed to acquire two Sharp 'singles' passenger locomotives in time for the opening between Cardiff and Navigation. He also ordered four 0-4-2 mixed traffic tender engines from W. Hawthorn of Newcastle-on-Tyne. The Sharp 'singles' arrived in September 1840, with Sharp also supplying engine drivers. 'Driver Gregory' arrived with the first engine named *Taff*, with Driver Estcourt following with the second engine named *Rhondda. Taff* was to haul the opening train on 8 October, but Bush had tried out the engine a week earlier when he had attained a speed of 60mph (the TVR's original Act of Parliament had a clause inserted to limit

Pontypridd in the 1850s. Here, the river Taff and its junction with the Rhondda can be seen. William Edwards' bridge of 1756, with the 1857 road bridge alongside, can be seen on the right. From the beginning, the TVR ran its trains on the right-hand side instead of the left, until a connection was made with the Newport, Abergavenny and Hereford Railway in 1858. (From Daffydd Morganwg, Hanes Morganwg, *1874)*

Close up of the 1850s lithograph. In this view, the railway, station and viaduct over the Rhondda River can be clearly seen along with disc and cross-bar signals, which are another feature of Brunel's railways. The tramroad built by Dr Griffiths, the original Rhondda coal promoter, can also be seen in the foreground. Walter Coffin's tramroad ran from his pit at Dinas and connected with Dr Griffiths's tramroad. The Rhondda branch was opened in 1841, and served by a short radius curve from the TVR on the north side of the Rhondda river. (From Daffydd Morganwg, Hanes Morganwg, *1874)*

speed to 12mph). Passenger services consisted of two trains each way every day including Sunday, trains leaving Cardiff at 8 a.m. and 3 p.m.. The return train left Navigation (Abercynon) one hour later in each case. The service was operated by one engine, the second Sharp 'single' being kept ready in case of a breakdown.

By the end of 1841, the locomotive stock comprised eight engines; *Taff*, *Rhondda*, *Merthyr*, *Cardiff*, *Llancaiach*, *Dinas*, *Dowlais* and *Plymouth*. *Dowlais* and *Llancaiach* were stationed at Merthyr to work trains above the incline.[13] In 1846 the TVR took delivery of *Newbridge*, the engine that had been at the centre of a dispute between Fisher and Colville when it was being tested. *Newbridge* was a 0-6-0 engine built by Benjamin Hick & Co. of Bolton in 1846. It has a historic place in railway history as it is possible that this photograph (at Cardiff docks) may have been taken in 1849 which, if positively confirmed, would make it the earliest known photograph of a railway locomotive. The *Newbridge* photograph came into the possession of the South Wales Police Museum through the grandson of the TVR policeman standing alongside the engine, P.C. John Wallbridge. The exact date of his move from the TVR police to the Glamorgan Constabulary is not known, which gives rise to uncertainty about the date of this photograph.[14] Most of the railway companies maintained their own police force, not only to patrol the railway and safeguard company property, but also to carry out signalling duties.[15]

Another breech of the company's rules was that of employees keeping a public house, an offence that the engine driver Thomas Newton was found guilty of in 1854. His fellow workmates, Samuel Thomas, Edward Rimron and John Powell, brakesmen and firemen, were found drinking in Newton's public house when they should have come on duty at 4 p.m. that Saturday. Their first task was to take charge of a pilot engine and take a train of coal down to the terminus and into the coal sidings. They should have relieved Edward Hopkins and others when their turn (shift) came to an end at 4 p.m., but when they did not show Hopkins went looking for them and discovered them at 6 p.m.; when he asked if they were coming on duty, he was greeted by a resounding 'No'. George Fisher apparently heard of this half-an-hour later, and hauled them in front of the Cardiff Police Magistrate, John Batchelor Esq., then Mayor of Cardiff.

Fisher, on the part of the TVR, charged the men with negligence of their duty; by failing to move the train and clear the line, they had endangered the lives of the public and the property of the company. The only real defence was put up by Newton who complained about not being paid his wages in the afternoon. Fisher replied that it was usual for the men not to be paid until Saturday night, but on application wages could be paid, as a favour, on Saturday mornings. It transpired that Newton had been paid at 4.30 p.m., but only £2 instead of his full wages of £5. Fisher replied that the reason Newton had not been paid his full wages was because '... he had been fined for being drunk on the East Branch.'[16] Newton also complained of working double turns and that if his relief did not show up on time he would act for him, and not a word was said on the subject before this occurrence:

Aerial view of Pontypridd, looking north. Brunel's Newbridge Viaduct can still be seen, with the station layout consisting of a long island platform dating back to the 1907 reconstruction. (SKJ collection)

'Men frequently act for each other: and in absenting himself on this occasion he had done no more than he had done no more than others did occasionally with impunity. He seemed to complain of undue severity towards him.'[17]

The other men also felt that they had been found guilty of an offence which was generally passed over. Batchelor animadverted[18] upon the conduct of the defendants and stated that he had the power to fine them each £10, or to imprison them with hard labour for two months. At this point Fisher said that he could not complain of the other three, '... but if Newton has been fined once he has been fined fifty times.'[19] Newton was fined £5 and costs, or six weeks' imprisonment with hard labour. Thomas and Rimron were each fined £3 and costs or one month's hard labour, while Powell was fined £2 and costs, or three weeks' hard labour. There was no reference to any of the men losing their jobs and, in the light of Fisher's comments, it would appear that the TVR did not want to lose their services, even if it had to fine them once or fifty times to keep them in line.

The TVR possessed three first-class, fourteen second-class and eight third-class carriages in 1841. It also had two horseboxes, forty goods trucks and wagons and two timber carriages, '... freighters find their own waggons upon this line'.[20] The regulations allowed coal and iron wagons provided they were sprung and had 3ft-diameter wrought-iron wheels to carry 5 tons each. Such an arrangement allowed engines to

Newbridge *locomotive, 1849-50. The mast of a ship in the west dock can be seen behind the engine. Also visible is the engine driver at the controls (all the original drivers are believed to have come from Newcastle), and the second railway policeman standing by the tender. PC Wallbridge is standing in the middle of the photograph. There was a Policeman's Box at Herbert Street Bridge, which faced the gasworks, to regulate the passage of down trains. (Courtesy of South Wales Police Museum)*

TVR Bridge over the connecting canal to the West Bute Dock at the top of Bute Road, Cardiff, 1973. Edward England's Potato Warehouse can be seen behind. (SKJ photograph)

take down trains of thirty wagons carrying 150 tons of coal. The third-class carriages had been delivered to the TVR at the end of October 1840 to allow a mixed train to be run on week days only, starting on Monday 2 November 1840. These third-class carriages were little more than goods wagons with wooden seats along the sides and the end. Entry was by means of a wooden flap which was lowered on to the platform in wagon style. Goods and animals would be carried in the carriage between the seats and there was no roof to the carriage. This is confirmed by W. Davies in his description of the TVR rolling stock, and travelling from the TVR station in Queen Street (then known as Crockherbtown) shortly after the line was opened:

> 'The Taff Vale Railway station in Queen Street in 1842, when I first passed through Cardiff to Trefforest, was a wooded structure. There was a bell fixed on two upright posts, which was rung the first time to warn intending passengers to prepare for the journey; rung the second time to come to the station; rung the third time to announce the train was going to start. The T.V.R. was then a single line. An open box truck was the second and third class passenger carriage, with an iron chain in the middle to divide the second and third class passengers. The second class had seats in rows, the third had only seats on the sides of the division.'[21]

The principal stations, Merthyr and Cardiff, appeared to have been designed by Bush, but no doubt under Brunel's direction, and both were equipped with train sheds. The intermediate stations of Llandaff, Pentyrch, Taff's Well, Newbridge and Navigation House (Abercynon), were described as being: '... executed in an economical style.'[22]

Llandaff station in the 1860s. On the right is the station master's house, office and booking hall, built in 1847. (Joseph Collings photograph/John Minnis collection)

George Bush had referred to the second-class passengers being troublesome, but it would appear that third-class passengers had even more reason to cause trouble, travelling, as Edwin F. Roberts put it, in the '… most detestable, disgusting, and misanthropic third-class carriage it was ever my ill-luck to travel in.' Roberts' journey was made in 1852, and the following account is his experience:

> 'At last we arrive at Newbridge, and putting up horse and car, wait for the train and take observations; for the eye is here insensibly attracted by the surrounding prospect. Situated in an amphitheatre of hills, with the Taff winding sinuously in the depth of the valley, one sees a fine grove of trees rising on the slope of a hill almost parallel with the descending road, out of which handsome villa looking houses pleasingly peep from their greenery. Across the town, where the Taff still indicates the valley, you behold the bridge of Pont-y-prydd, with its single circular arch of a light and lofty span, leaping with airy boldness across the stream. Right and left the new town stretches, sheltered on every side by declivities which are tinted with that rich autumnal green so pleasing to the eye. A lofty height right over the station invited me to ascend, but the warning whistle of the advancing train prevented me from following the irresistible impulse. In a few moments we were shut in, and were whirled away in the most detestable, disgusting, and misanthropic third-class carriage it was every my ill-luck to travel in. It seemed as if the projectors of the railway had a selfish desire to deprive the passenger of all the external beauties on either hand, and it was only by rushing from one small opening (about a foot square) to another that I could see anything of the locality I was being swept through at all. The doors open inwardly, and as the guards are not over nice – (one irascible official in particular

> always had a threat of expulsion on complaint) - one's legs are in absolute danger, if necessity or a choice places you near the corners - of being assaulted. Since I am on the subject, I will mention a case of gross neglect of the rights of passengers, and so ease my conscience on the matter. It was on a Saturday evening that I returned, and at almost every station there were relays of workmen, navigators, and country-people, male and female, waiting to be taken up by this, the last train. They amounted to hundreds in all; but, as they mostly went but short distances, that which would have been impossible, by the most refined skill in packing up, was rendered possible on the carpet-bag-packing principle. Still journeying by the third-class, the carriage became filled, legally, but uncomfortably filled. At the next station good-nature did not object to be discommoded a little. At every fresh station more came in than went out, until every square inch of the floor was used for standing-room, and the smell and closeness of the atmosphere - thick, "slab," and rich enough, to one not over –particular - became almost unbearable. At a further station I stoutly remonstrated with the guard, and was advised to be silent or I should be "pulled out pretty quick, dam me." Finding that I had to deal with a gentleman who felt the full impress of the "insolence of office" I took the initiative and got out at once, declining to journey farther in a carriage that held, at the least, double the number it ought. There were some half-a-score of brawny, beery, "navvies," ready to jump in; but, finding that the absence of one would not make room for three, they very noisily followed a motion made to "get into a tub" - the said "tub" being a stone, coal, or iron car, very open indeed, and without any lining to its seatless sides. My friend in blue and buttons finding me intractable, and having put it out of his power to "pull" me out, was fain to give me a seat in a second-class carriage; and several women, who had been mercilessly crushed by the crowd and annoyed by the lax freedom of the aesthetic converse somewhat loudly carried on, were accommodated in the same way. Two or three additional carriages, which were absolutely necessary and which lay in idle plenty, everywhere at hand, would have obviated all the difficulty, and remedied the scandalous annoyance. It was fortunate that I had taken my ticket as also possession, thus entering into a contract, or I should have been left behind - a state of unprotected orphanage that "can be better imagined than described.'[23]

Of the 2,120 rules in the 1853 TVR *Rules and Regulations Book*, a number were exclusively for those parts of the line where special conditions applied, such as working the main incline at Aberdare Junction (Abercynon), which accounted for ninety-eight rules (Nos 1735 to 1833). The incline at Llancaiach needed ninety-nine rules (Nos 1834 to 1933), which included the rule that all chimneys of engines working on this branch had to be lowered (it is spelt out that the chimney has to be lowered before and while passing under the bridge) '... over the Inclined Plane, and the Tramroad Bridge, near the Nelson Public House.' (No.1932). The Dowlais incline, however, was managed with sixty-one rules (Nos 1933 to 1994), while the Llantwit branch at Maesmawr got by with only nineteen rules (Nos 1995 to 2014).

This Llantwit branch was a short branch from the main line to the Dihewyd or Llantwit colliery, owned by Thomas Powell. Its origins can be seen as a branch line

Llandaff station, photographed during demolition, 1972. (SKJ photograph)

proposed as part of George Overton's proposal to extend the line of the Merthyr Tramroad down to Cardiff in 1823. The idea of the Llantwit branch was revived several times, most significantly in the wake of the Taff Vale Railway's Act of Incorporation, when the engineer W.H. Harrison surveyed a number of lines for the Marquess of Bute. Although this covered lines to Llantrisant and St John's chapel near Tonyrefail, with extensions into the Rhondda and the Ely Valleys, Harrison only recommended that the first 4 miles from Maesmawr should be undertaken initially.[24] Around this time (1837), Powell was to take an interest in the proposal due to his acquisition of mineral rights in the locality. By 1842, Powell's railway was to take shape as the Llantwit Vardre Railway, and was opened on 28 December 1843. Powell had Harrison's line resurveyed, and reduced the costs of working the 297-yard long incline plane, which had a gradient of 1 in 6.6, by operating it on the balance-load principle (self-acting).[25] Connection to the TVR came into affect on 25 April 1844, and a train carrying Powell's coal from his Llantwit colliery was worked through to Cardiff.

Neither Brunel nor the TVR had any input into the design or construction of the Llantwit Vardre Railway and, for operating purposes, it had an independent existence until it was acquired by the Llantrissant & Taff Vale Junction Railway in 1861. It was Powell's colliery interests at Gelligaer that had been instrumental in the TVR constructing the Llancaiach branch, which also had an inclined plane. Despite being a director of the TVR, Powell continued to use the canal to ship much of his coal and was frequently in dispute with his fellow directors over shipping rates. An extra set of working instructions were put in place for the main incline at Abercynon in 1865-66, when it was worked, without the assistance of cable haulage, by locomotives alone. The Colliery Guardian of 10 November 1866 [26] not only records this fact, which has largely gone unnoticed, but gives a brief description of the working of the main incline under cable haulage:

'The principal feature of the Taff Vale Railway is a steep incline of one in eighteen, and half a mile long, from Aberdare Junction to the top of a range of hills leading to Merthyr, From Cardiff the line rises throughout its entire length, but in the course of its ascent to the mountainous country of the north of Glamorganshire, the enormous mineral and passenger traffic has had to be carried, locomotives and all, up and down the Aberdare Junction incline by means of ropes and a stationary engine at the top of the hill. The rope was coiled on and off a circular drum, with its axle crossing the railway, and supported on each side by two substantial towers of masonry.

This incline, with its ropes and drums, is as well known and famous in Wales as the rope travelling of the Blackwall Railway to the older inhabitants of London. For years the contrivance of pulling the trains up into the Welsh mountains worked well without any accident of note, except upon one occasion when the axle broke and let a mineral train down with a run, without doing further damage.'

The comment that the incline was '... well known and famous' is probably no exaggeration as earlier visitors such as Michael Faraday had noted with great interest the working of a self-acting incline on the Dowlais tramroad system in 1819.[27] Nicholas Wood describes the theory of such inclined planes and those assisted by steam power: '... the introduction of steam engines, to drag carriages up ascending planes, upon railroads, is comparatively recent.'[28] Wood goes on to attribute the first example of this kind to Mr S. Cooke, who applied an engine to draw up wagons of the Urpeth colliery at Birtley Fell in the county of Durham in 1808. Brunel himself regarded the main line incline as an example worthy of attention, and in 1843 he sent Benjamin Herschel Babbage, his assistant engineer on the Genoa railway project and the son of Charles Babbage, to study the inclined planes and other works on the line.[29] To return to the Colliery Guardian's description of the main line incline:

'In working the trains up and down this incline the plan adopted was never to attach a weight of more than thirty tons to either end of the wire rope, the long mineral trains being divided before making the descent, and a sufficient number of wagons always kept at the top of the incline to counterpoise to some extent those to be drawn up. The short passenger trains of the line were always drawn up without being separated and without delay, but four wagons of iron or iron ore were the utmost the rope was allowed to bear, and to these about ten empty wagons would form a good working balance, although the steam power applied to the drums enabled very unequal weights at the two ends of the rope to be used.

The fact that all the minerals and iron of Merthyr had to be carried down to the sea caused the daily downward traffic on the incline to be more than double the weight of that brought up, unless in a few exceptional instances, so that, as a rule, very little work was required from the stationary engine.'

It was not all downhill as far as operating the incline was concerned for there were problems when more wagons than that allowed were attached, as in 1843 when George Fisher was forced to write to Thomas Evans of the Dowlais Iron Co.:

'... I beg to inform you that Five of your Waggons were damaged this morning at the Main Incline, one of the bank riders placing on the rope more Waggons than his instructions allows him to do, and the rope broke. When they came to the bottom the road was clear for them, but one of the Six wheeled wagons broke the crossing plate and they ran off the Rails...'[30]

The Dowlais Iron Co. had been criticised the year before by the secretary of the TVR, William Burgess, on the actual condition of their wagons. It was not the first time either, '... I have not troubled you lately on the subject of your wagons having understood that you had determined on a Radical change...'[31] This referred back to a report by Burgess read to the TVR directors in their meeting of 24 May 1842, on the condition of the Dowlais Co.'s wagons; '... So defective a State as to cause constant and imminent danger to human life as well as great loss by breakage.'[32] This was despite a meeting with Burgess and a representative of the Dowlais a month later on the 'particular evils' of the form of couplings used on their wagons, and further comment on their '... wretched state of repair, not more than about half of them at all fit for work...'[33] Despite the problems of wagons, the main incline continued to be used for the next twenty-five years, and was capable of being used for longer but for the increase of traffic going up the incline: 'Much mineral produce from the Rhondda Valley has had, during the last two years, to be carried up the incline, so that the economy of power is not now so great as in former times.'[34]

Morganstown crossing in the 1950s. The line of the Melingriffith tramroad (a works branch railway in this photograph) across the TVR can be clearly seen. The GWR 5600-class 0-6-2T engine is returning empty colliery wagons. Note the four lines at this point and the loading gauge on the right for access to the works branch. (SKJ collection)

The *Colliery Guardian* highlighted the fuel savings (for the stationary engine) and the reduced running costs that had been due to the main flow of traffic being in a downhill direction. This position changed when coal from the Rhondda and Cynon Valley was being routed up the incline and through Quakers Yard to inland destinations. It accelerated the need to replace the main incline with a locomotive incline. Replacing the main incline as an operational improvement had been considered by the TVR as early as 1853 and, as a measure of forward planning, it included this option in its Act of 1857, '... to construct new Lines of Railway, to alter, widen and improve Portions of their existing Railway, and for other Purposes.'[35] It was proposed to replace the main incline with a locomotive incline having a gradient of 1 in 36, but the powers were not intended to be used immediately; indeed in February 1859, George Fisher reported to the directors that the replacement of the main incline would not be financially viable.[36] By 1864, for the reasons mentioned, the impact of the sharp increase in traffic now using the incline as well as the increased costs for working, caused a change of attitude. In March 1864 the replacement of the main incline was recommended by George Fisher and adopted by the TVR board on 18 March 1864. There were to be some changes; the new incline was to be built with a gradient of 1 in 40 instead of the 1 in 36 originally proposed in the 1857 Act. This was to result in heavier earthworks but, before the new diversionary line was actually opened, the main incline was to be operated by locomotives alone:

> 'The abolition of this heavy incline having been decided upon by the company, a second one, a quarter of a mile longer than the first, with a gradient of one in forty, was commenced more than a year ago, and will be finished in about six months. The new incline lies parallel with the old one, but in a cutting seventy feet in depth for a considerable portion of its length. The soil being hard, and full of large fragments of rock, most of the work has been done by blasting. Soon after the works were commenced the company removed the drums and stationary engine, and for the last twelve months have carried all the trains up and down a single line on the old incline of one in eighteen by the aid of two or three locomotives. As this is by far the steepest gradient in the kingdom over which passenger trains and heavy mineral traffic of a main line are so worked, very stringent regulations have necessarily been made to ensure safety.'[37]

The working of the main incline by locomotive power during this period was a significant feat. If it was not for the work actually being started, one wonders if it was an attempt by the TVR to see if a new incline was actually needed, especially as a section of the original line at the top of the incline was widened in 1863, only to be abandoned, like the rest of the main incline in 1867. In August 1867 Fisher reported that the new locomotive incline was completed and now in use.

The Llancaiach branch, which included another inclined plane, opened in November 1841 as a branch line off the main line at Stormstown Junction. The line crossed the river Taff on a four-arched viaduct. Stormstown took its name from one

Llancaiach Viaduct, looking towards Llancaiach in 1985. Note the shelf of the top of the incline cutting into the hillside. (SKJ photograph)

of the original contractors of the Llancaiach branch, Storm and Douglas. They, however, failed to complete the works, and the contract was taken over by John Calvert. The branch was 3 miles 29 chains long and included a self-acting incline at a gradient of 1 in 8½. The incline ceased to be used after 3 March 1870, when traffic was diverted via Quakers Yard Junction and Aberdare Junction on the Newport, Abergavenny & Hereford Railway's Taff Vale Extension line. A passenger service operated from Pontypridd to Nelson via Cilfynydd was not available until 1 June 1900 (the original Llancaiach branch was for mineral traffic only). Much of the incline embankment was to be removed to accommodate the construction of the A470 trunk road in the 1970s; the further construction from Abercynon to Merthyr in the 1980s was to destroy remnants of the TVR main incline. It is still possible to walk part of the track bed at the summit, which can be seen as a 'shelf' cutting into the hillside. The viaduct itself was removed following the aftermath of the Miners' Strike of 1984/85, when access to Abercynon colliery (built 1889-91) was no longer required.[38]

On the subject of travelling on the TVR, second-class and first-class carriages were typical of the period, with separate compartments which were not inter-connected. Smoking was strictly prohibited by the TVR in carriages and even on the platform, and was covered by TVR general regulation No.50: 'Smoking by any

River view of the Llancaiach Viaduct in 1985, with the access bridge to Abercynon Colliery behind. (SKJ photograph)

Passenger… upon any part of the Company's premises, or in any Carriage or Train, is strictly forbidden, whether such person be travelling, or otherwise.' Brunel gave an account of travelling on the TVR, with particular reference as to how he tried to avoid that particular regulation, to Richard Price Williams (1827-1916). Price Williams, an engineer who was related to Walter Coffin, recalled the incident in which Brunel told the story of his experience with the 'incorruptible guard, Meyrick' at Dowlais station (photograph on p.197):

> 'I had been up to Dowlais and was quietly smoking a cigar on the platform waiting for the train back to Cardiff when, Meyrick very respect-fully came and reminded me that smoking was strictly prohibited on the platform. I strolled out into the Goods Yard, as there was some little time to wait before the train started, and on my return I requested Meyrick to reserve a compartment for me, which he did. When well out of the station, and as I was comfortably enjoying a smoke, to my surprise there was a tapping on the window, and Meyrick's face appeared. In the most sorrowful tones he said, "There you are, at it again, Mr. Brunel, a-breaking the rules". I at once let down the window and handed him a tip, but Meyrick in a still more saddened voice, exclaimed, "Oh! Mr. Brunel, you don't surely think so badly of me as that. I am only doing my duty."'[39]

Above: *Pontypridd station, 1879. The Brunel style station buildings with overhanging roofs providing platform cover. Courtesy of Pontypridd Reference Library.*

Dowlais works and railway station, c.1870. Here, the photographic reality of the Dowlais landscape can be contrasted with Penry Williams' impression of the Cyfarthfa works. In the span of some forty years between both scenes, Dowlais had emerged as the most progressive of all the Merthyr ironworks and a pioneer of the steel era. Dowlais railway station can be seen on the left. This line was originally planned as a branch of the TVR but was eventually constructed by the Dowlais Iron Co., with an inclined plane of 1 in 12 from a junction with the TVR, to the lower entrance of the Dowlais works. The 'Little Rail Mill' dominates the foreground. Dowlais had supplied rails for the GWR which Brunel praised as being; '... some of the best we have ever had.' (John A. Owen collection)

Opposite below: *Taffs Well, mid-1860s. The station building, seen beyond the road bridge, dates back to 1847 when it replaced a timber structure. It was closed in 1863 when Walnut Tree Junction was opened. (Joseph Collings photograph/Stephen Rowson collection)*

Chapter 10 Notes

1 *The Great Western Railway Magazine*, 1908, 'Some Reminiscences of Brunel' by R. Price Williams.
2 PRO RAIL Letter from Francis Fox to George Fisher, 17 September 1849. Francis Fox was later to become the Engineer for the Bristol & Exeter Railway, in which capacity he was responsible for the design of the single-span curved roof of the B&ER's through train station at Temple Meads, completed in 1877. See Binding, John, (2001), p.146, *Brunel's Bristol Temple Meads*. Oxford Publishing: Hersham.
3 Cardiff City Reference Library, TVR Book of Rules and Regulations 1853, L656.2.05.
4 PRO RAIL 684.2, TVR Director's Minute Book. See also Chapman, Colin, *A Clear Road?*, in the *Welsh Railways Archive*, Vol.1, No.3, May 1991.
5 Symons, Jelinger, (1854), *The Industrial Capacities of South Wales*, published in the *Cambrian Journal*, Vol.1, 1854.
6 *The Railway Times*, pp.137-38. 5 February 1842. This is based on the second or unpublished volume of 'The Railways of Great Britain and Ireland' by Francis Whishaw Esq., Civil Engineer, M. Inst. C.E.
7 *The Railway Times*, p.138
8 Matthews, John Hobson, (1898-1911), pp.327-28, Vol.5, *Cardiff Records: being materials for a history of the county borough*, Cardiff Corporation: Cardiff. Extract from a letter of Mr W. Davies, 15 March 1899.
9 *The Cambrian* 26 September 1841.
10 See chapter 7.
11 Highton, Edward C.E., (1852), *The Electric Telegraph : Its History and Progress.* John Weale: London.
12 Chapman, Colin, (1997), p.143, *The Nelson and Ynysybwl Branches of the Taff Vale Railway.* Oakwood Press: Headington.
13 The practice of naming engines lasted until 1863. Barrie, D S M. (1969 ed. originally pub.1939), p.33.
14 In the absence of an exact date, this distinction is currently attributed to an 1850 photograph of a Norris locomotive in Philadelphia, USA. Interestingly a Norris locomotive found its way to the TVR in the shape of the first passenger locomotive to work the Aberdare Railway.
15 A studio portrait of a TVR policeman; PC Ambrose William Pontin, resplendent in his dark blue frockcoat, can be found in John Hutton's book; *Taff Vale Railway Miscellany*. Hutton, John. (1988), *Taff Vale Railway Miscellany*. Sparkford: Haynes Publishing Group. A FOULIS-OPC Railway Book. Details on TVR uniforms, using PRO RAIL 684/94 records, can be found in the Glamorgan family History Society Journal No.27 1992. Muir, Jill. *The Taff Vale Railway Company 1840-1924*. pp.8-9.
16 *Cardiff and Merthyr Guardian, Glamorgan, Monmouth and Brecon Gazette*, 14 January 1854.
17 *Cardiff and Merthyr Guardian, Glamorgan, Monmouth and Brecon Gazette*, 14 January 1854.
18 The expression, animadverted, means to pass criticism or censure on conduct or fault etc, to literally give someone a piece of your mind.
19 *Cardiff and Merthyr Guardian, Glamorgan, Monmouth and Brecon Gazette*,14 January 1854.
20 *The Railway Times*, p.138.
21 Matthews, John Hobson, (1898-1911), p.327. Extract from a letter of Mr W. Davies, 15 March 1899.

22 *The Railway Times*, p.138. Quoting from Francis Whishaw's unpublished work. C.E. Bernard, who later become a prominent architect in Cardiff, appears as one of the principal shareholders in the TVR's first listing, but no evidence has come to light that he was involved in an architectural capacity on the TVR.

23 Roberts, Edwin F., (1853), *A Visit to the Ironworks and Environs of Merthyr Tydfil in 1852*, William Edward Painter: London.

24 Chapman, Colin, (1996), pp.10-13. *The Llantrisant Branches of the Taff Vale Railway*. Oakwood Press: Headington.

25 Harrison's proposals included a stationary engine to assist working of the incline, Chapman, Colin, (1996), pp.10-13. The Llantrisant Branches of the Taff Vale Railway. Oakwood Press: Headington.

26 Colliery Guardian, 10 November 1866, reproduced with additional comments by Colin Chapman in the *Welsh Railways Archive*, Vol.III No.8, November 2003, p.181.

27 Tomos, Dafydd, (1972), p.26. *Michael Faraday in Wales including Faraday's Journal of his Tour through Wales in 1819*. Gwasg Gee. See Chapter 4.

28 Wood, Nicholas, (originally published 1825, third edition 1838), p.251. *A Practical Treatise on Railroads*. Longman, Orme, Brown, Green &, Longmans: London.

29 PLB, 2C, pp. 151-61, 20 March 1843, IKB to W. Coffin, see Buchanan, R.A., (2002), p.90. *Brunel The Life and Times of Isambard Kingdom Brunel*. Hambleton and London: London.

30 GRO, 1843(1) f.470, George Fisher to Thomas Evans, 22 May 1843.

31 GRO, 1842(1) ff. 212,213, William Burgess to Sir John Guest, 3 November 1842.

32 TVR Directors' Minutes, 24 May 1842 and GRO 1842(1) f.204.

33 GRO 1842(1) ff. 205, 206. 25 June 1842.

34 *Colliery Guardian*, 10 November 1866, reproduced with additional comments by Colin Chapman in The Welsh Railways Archive, Vol. III No. 8, November 2003, p.181.

35 Taff Vale Railway Act of 1857, cliii (L. and P.)

36 Chapman, Colin, *The Main Incline 1841-1867*, in the *Welsh Railways Archive*, Vol.1, No.3, May 1991, pp.78-80.

37 *Colliery Guardian*, 10 November 1866, reproduced with additional comments by Colin Chapman in the *Welsh Railways Archive*, Vol.III No.8, November 2003, p.181.

38 The mineral traffic was actually carried on a separate steel girder bridge which can be seen, in the photograph, through the arches of the viaduct, the opportunity taken at the time to improve the road access under the road arch of the viaduct.

39 *GWR Magazine*, 1908, article by R. Price Williams. Meyrick was referring to regulation No.42; 'Not any person is allowed to receive any gratuity from the public on pain of dismissal' although it was to be found in a number of other general regulations i.e. No.53.

11
Beyond Brunel
'When the Coal Comes From the Rhondda'[1]

Following the opening throughout of the line, the TVR was to experience a period of financial uncertainty which severely depressed the price of its shares. As a Welsh railway, its fortunes were to be transformed with the exploitation of the coal seams of the valleys it served, and it was from this base that the TVR was to develop into one of the most efficient and prosperous railways in the country. As a Brunel line it was a unique in several ways, the most obvious difference being Brunel's one and only deviation from the broad gauge to the narrow or standard gauge in Britain. Largely because of this, it had no real ties to the growing empire that the GWR was to become, and the ties to Brunel diminished over the years. Brunel was to have some, but not a great deal, of contact following the opening of the TVR main line, and he appears to have made no comment on one of its early acquisitions, the Aberdare Railway in November 1846.

Three years after the opening throughout of the TVR, proposals for building a railway that would continue the line of railway communications into the Cynon Valley began to be mooted.[2] The river Cynon flows into the Taff at Navigation House (Abercynon) but the line was to take its name from the settlement that had grown up around the basin, formed where the tributary of the Dare joins the Cynon, hence Aberdare. In 1844 local ironmasters and colliery proprietors, led by Crawshay Bailey and the TVR's chairman, John Guest, decided to undertake a survey of a route for such a line. As the name suggests, Crawshay Bailey was related to the Crawshays, being a nephew of Richard Crawshay. With his brother Joseph, he had learnt the art of ironmaking at Cyfarthfa. By 1820 he was a partner in Nant-y-glo, which his brother Joseph and Matthew Wayne had purchased in 1811.[3] Crawshay Bailey was responsible for the acquisition of the nearby Beaufort ironworks in 1833, to feed the Nant-y-glo forges and mills with pig iron that was converted into rails. In the 1830s and 1840s, J & C Bailey were one of the largest suppliers to the growing home and American railway market.[4] Crawshay Bailey's interest in railways was to make him the butt of local doggerel, to be immortalised in lines such the following:

'Crawshay Bailey had an engine
And he found it wouldn't go
So he pulled it by a string
All the way to Nant-y-glo...'[5]

Following the death of Anthony Bacon (the eldest son of the Cyfarthfa founder) in 1836, Crawshay Bailey acquired his Aberaman estate in the Cynon valley.[6] He took up residence in Aberaman House but did not exploit the mineral wealth of the estate until 1845, when he erected his first furnaces there. In the same way that the Glamorganshire Canal's transport monopoly was challenged by the new railway, the Aberdare Canal was also seen as a past mode of transport, particularly in sending down coal from the growing number of Cynon valley pits. A surveyor who was later to take on the role of engineer, David Jones, was appointed in 1844 to make a survey of a line from a junction with the TVR, referred to in the resulting Act as Ynys Meyrick.[7] From this junction near the foot of its main incline at Navigation House (Abercynon), it would proceed up the Cynon valley to terminate near the Gadlys ironworks at Aberdare.[8] David Jones began his survey on 17 August 1844 and plans, checked by a civil engineer, E. Scott Barber, were made in time for a Bill to be put before Parliament in November 1844.

On 31 July 1845, the Aberdare Railway gained its Act of incorporation which also authorised the leasing of the railway to the TVR, subject to the approval of three-fifths of the proprietors.[9] John Guest was to become its first chairman, with Crawshay Bailey as treasurer. As an undertaking, it was clear that the Aberdare Railway would be in the TVR's orbit from the start, the question being how long would it have an independent life? In October 1845, the Aberdare proprietors first raised the suggestion that the TVR should take over the undertaking, almost before any construction had started. Some time later, and after much internal dissention, the TVR finally agreed to take over the Aberdare Railway, the lease coming into effect from 1 January 1847, which was less than six months after the opening of the line. The line posed no major engineering challenges and was constructed relatively cheaply, the cost per mile being just over £4,000, compared to almost £20,000 per mile that the TVR cost.[10] The implications of such cheap construction were soon to become evident. Future problems and high running costs resulted in a line that suffered from severe curvature, as it closely followed the river banks and had adverse gradients. One of the issues raised against the call for leasing the Aberdare Railway at the TVR's General Meeting of 15 December 1846 concerned the unfavourable curves and gradients. David Jones, the engineer of Aberdare Railway, sought to refute the charges that the line was causing premature wear and tear, particularly of locomotives bearings.[11]

The line was being worked by the TVR on 1 January 1847 and its short, and somewhat limited, independent history was over. It was to bring with it substantial trade for the TVR in carrying the steam coal raised in the Cynon Valley to Cardiff. Brunel was not involved in this phase, at least not on the side of the TVR, as he was later instrumental in tapping the higher reaches of the valley and its steam coal through a branch of the Vale of Neath Railway. Indeed, George Fisher was to be caught out by Brunel as Fisher had advised the TVR directors that the Dare valley would not be a coal mining district of any significance, and that its future was in iron-working.[12] In the event, Brunel's broad-gauge Aberdare Valley Railway was to come in and exploit the coal reserves of an area that never took off as an iron centre. All this was in the future and another story, the acquisition of the Aberdare Railway, was to help transform the fortunes of the TVR; indeed, the story of the coal exploitation of the Cynon Valley involved a number of names that were to be prominent in the development of the Rhondda coal trade, including coal-owners like David Davis, who moved on from Abercwm-boi to sink the first of the Ferndale pits in the Rhondda, and Samuel Thomas (father of D.A. Thomas), who moved on from Bwllfa to Clydach Vale.[13]

The depression and financial constraints that hung over the TVR in its early years reached such levels that it could not afford to pay its contractors. Charles Wilkins relates the story of Howells the contractor who, with his partner Evans, pressed the TVR for payment of their completed contract which included the construction of the Goitre Coed Viaduct. Walter Coffin, then chairman, said: '… we have no money; we cannot pay.'[14] Howells and Evans were to settle with payment in shares at the market value; this must have appeared to be a considerable risk at the time, but as there appeared to be no other form of remuneration available, it was taken. In a few years it proved to be the greatest bargain they had ever struck.

In his original report, Brunel estimates the projected revenue of the railway at £4,076 16*s*. Again, Wilkins takes pleasure in highlighting that the revenue taken in the last year (1887) was: '… over ninety two thousand pounds.'[15] This is obviously a criticism of Brunel in that he failed to take into account the potential of the coal and the impact it would have on the trade of the railway. However, Brunel's involvement was at the behest of the ironmasters of the Dowlais, Penydarren and Plymouth works, successors to the original promoters of the Merthyr Tramroad, and the overriding objective of the TVR, in continuation from the Merthyr Tramroad and the Glamorganshire Canal, was to carry iron from Merthyr to Cardiff. In the second report, some coal traffic is noted, mainly from the Waun Wyllt collieries and the Bargoed valley, but there is little recognition of potential for trade posed by coal, even though 113,749 tons of coal had been sent down the canal during the year 1830, compared with a total of 87,367 tons of iron. That this was not pushed further seems to be at odds with the fact that Coffin, with colliery interests in the Rhondda valley, became the vice chairman of the TVR, and that a leading director was Thomas Powell (1784-1864) of the Gellygaer colliery, near Llancaiach. Both industrialists, however, had their own transport interests which did not always coincide with the interests of the TVR. What was to become patently clear in the years after the completion of the TVR was that coal was to have a major impact, and affected not only the financial prosperity future of the TVR, but had a considerable impact on engineering aspects. The great increase in traffic caused the line to be doubled from single track, with passing places, to a double main line sooner than anticipated, with separate up and down mineral lines built between Cardiff and Pontypridd, along with several large sidings to accommodate the vast quantities of coal traffic awaiting shipment. All of these works were completed by the mid-1880s.

There was another engineering link, both in terms of Brunel's original proposals for the TVR and with Brunel playing a part. This was to come about in the mid-1850s when a number of the major traders on the TVR began to raise concerns about the Bute trustees' dominant position and the impact it could have on their business with regard to shipping through Cardiff. There was a reluctance by the Bute trustees to allow the TVR access to the new Bute East Dock and consequentially, a number of prominent traders, all connected with the coal trade on the TVR, began to discuss the ramifications of an alternative outlet to the sea, in essence reviving the original TVR scheme of a dock or tidal staithes on the river Ely. In 1855 this powerful business group deposited a Bill in Parliament which became law as the Ely Tidal Harbour & Railway Act, on 21 July 1856. Naturally, there was very strong opposition from the trustees. The Act authorised the construction of a tidal harbour on the eastern bank of the river Ely, together with a railway around 6¼ miles in length, connecting the harbour with the Taff main line by a junction in the parish of Radyr.[16] The group included: the Hon.Robert Windsor Clive (owner of mineral rights and later to become involved with the Barry Railway proposal); Crawshay Bailey; Thomas Powell; Revd George Thomas; William Cartwright; James Insole; Thomas Wayne and John Nixon. Even though some of them were TVR directors or shareholders, they were to act independently of the company in the interests of coal trade. There was also a major reason for them to distance themselves from official TVR policy i.e. the agreements entered into by the TVR over the Bute Dock. Therefore, the TVR did not oppose the Bill in Parliament, and other leading personalities associated with the TVR raised issues relating to the loss of traffic receipts the company faced. A second Bill was deposited in November 1856 to construct an enclosed dock near the opposite bank of the Ely to the tidal harbour. The name of the company was also to be changed to the 'Penarth Harbour, Dock & Railway' (PHDR) under the Act granted on 27 July 1857. The Act authorised the construction of the dock and a branch railway, which was to run from a junction (Grangetown Junction) with the railway authorised under the first Act, crossing the river Ely to reach the new dock (just under 2 miles in length). It would also have running powers over the whole of the TVR.

In both Acts, provision was made for working arrangements with the TVR or an arrangement for the TVR to work the line for a period not exceeding 10 years. By the time the railway and the shipping facilities were open for business on 4 July 1859, no such arrangements had been made, and the PHDR worked the line themselves with their engines. Working was largely confined to working the coal trains from the junction at Radyr, officially known as Penarth Junction, to the tidal harbour, and returning empty wagons to the junction rather than exercising their powers over the whole of the TVR. A Head of Agreement was finally reached in August 1862, with the PHDR to be leased to the TVR for a period of 999 years. The Bute trustees felt that this lease went against the agreements the TVR had entered into from the mid-1840s onwards when seeking access to the West Dock, namely to abandon the construction of a tidal harbour on the river Ely. The Bute trustees applied to the

Court of Chancery the same month, seeking to restrain the TVR from entering into such a lease. They failed in this action and appealed to the House of Lords in February 1863 to reverse the verdict of the lower court, but once again the verdict went against them. The PHDR had to obtain another Act to authorise the period of the lease (different from that authorised in either the 1856 or 1857 Acts).[17]

John (later Sir John) Hawkshaw was the engineer for this work and one of his assistants at the time was a young engineer, Henry Marc Brunel. Brunel's youngest son was originally apprenticed to William Armstrong, as a premium apprentice, for two years in October 1861. In October 1863 he continued his apprenticeship as a pupil with Sir John Hawkshaw, in time to be employed on some of the dock works at Penarth. Henry Marc records that in the first few months he was engaged in the drawing office on various projects including '... the Charing Cross Railway works...'[18] He goes on to state that:

> 'In June 1864 I, for about 3 weeks was at Cardiff taking copies of drawings of Penarth Docks then in progress... In August 1864 I went to Penarth and remained there till October 1865 – During this time the Penarth Docks were being completed and the docks were opened in June 1865 – I had charge under Mr. Samuel Dobson of important parts of the works and in his absence superintended on his behalf the working of the Dock for 3 or 4 months.'[19]

Cardiff Pier Head, c.1910. Steamtugs Eagle *and* Sylph *are at the entrance to the docks, where a vessel is leaving. (Photograph by H.J.B. Willis, SKJ/SR collection)*

Penarth Dock, shortly before its redevelopment as a marina in the early 1980s. Looking across the basin to the main dock where the coaling appliances were situated. (SKJ photograph)

Hawkshaw was involved with a number of projects associated with the Brunels, from the adaptation of Sir Marc Isambard Brunel's Thames Tunnel to railway use as part of the East London Railway to completing the Clifton Suspension Bridge (with W.H. Barlow) using the chains of Hungerford Suspension Bridge demolished during the construction of the Charing Cross Railway Bridge. He was also involved with the widening of the TVR main line above Navigation, and on TVR work in general from the late 1850s, but Henry Marc does not refer to this. A number of Hawkshaw's signed contract drawings are still used today by Network Rail engineers on former TVR lines, with some dating back to 1858.[20] An observation of Hawkshaw's engineering approach can be seen in a story told by the engineer Richard Price Williams (mentioned in Chapter 10), who sought to persuade railway companies to take up the use of steel rails. In this he initially faced an uphill task, particularly when he first tried to persuade Hawkshaw to adopt them on his railways. Despite the longer wearing attributes of steel rails, Hawkshaw exclaimed, 'What put down a brittle material like steel on the running road of our main line? … I should be tried for manslaughter.'[21] He changed his mind following the results of trials and subsequently told Price Williams, 'Now you can put down as many Bessemer rails as you like.'

Through his connection with Hawkshaw, Henry Marc was to be involved in completing part of his late father's plans for the Taff Vale Railway, the Ely Harbour branch – albeit in modified form. With the lease on the PDHR, the TVR became the first railway company to operate both dock facilities and a railway to the coal-

A busy day at Penarth Junction, Radyr, photographed by H.J.B. Wills in 1907. (SKJ/SR collection)

fields, setting the standard for future railway companies, such as the Barry Railway, to aim at. Henry Marc Brunel was to return to South Wales in connection with the latter, but that is outside the scope of this volume as is the association with one of Brunel's ships, the SS *Great Britain*, whose last commercial voyage was from Penarth dock.

While coal was to be the dominant factor in the TVR's future prosperity, it is interesting to follow the original *raison d'être* for the line i.e. carrying Merthyr iron to Cardiff, and taking it to its ultimate conclusion through Brunel's connection with George Thomas Clark (1809-1898). The connection between the two has already been shown; the link with the Guests, however, was actually made through the Conybeare family before the first recorded meeting with Brunel.[22] Clark, who had trained as a surgeon, was to work for Brunel as assistant engineer in charge of two major bridges on the GWR. Responding to Isambard Brunel's request for information on his father in 1870, Clark recalls the circumstances of his first meeting with Brunel during the winter of 1834/35:

> 'I made your father's acquaintance, rather characteristically, in an unfinished tunnel [Staple Hill Tunnel] of the Coal-pit Heath Railway ; and when the shaft in which we were suspended cracked and seemed about to give way, I well remember the coolness with which he insisted upon completing the observations he came to make. Shortly afterwards I became, at his request, his assistant; and during the parliamentary struggle

> of 1835, and the subsequent organisation of staff, and commencement of the works of the Great Western, I saw him for many hours daily, both in his office and in the field, travelled much with him, and joined him in the very moderate recreation he allowed himself.'[23]

In the same response, Clark commented on Brunel's temperament and working practices:

> '... He possessed a very fine temper, and was always ready to check differences between those about him, and to put a pleasant construction upon any apparent neglect or offence...Everything for which he was responsible he insisted upon doing for himself.'

Brian Ll. James, the editor of Clark's biography, points out that there was no doubt that Clark admired Brunel in terms of his ability as an engineer, his dealings with contractors, landowners and his own staff, his legendry appetite for work and his general demeanour, but that he was conscious that his authoritarian style was '... perhaps a weakness'. Clark recognised that his approach denied any independence of action in his subordinates, '... though he trained several excellent assistants, few of them ever rose to be engineers-in-chief, or attempted to any great or original

George Thomas Clark, 1809-1898. (SKJ collection)

Lady Charlotte Schreiber (Guest) addressing schoolchildren at Dowlais School in September 1855. George Thomas Clark is the larger of the two men in the front row. (Courtesy of Glamorgan Record Office)

work.' If Clark did look to Brunel as a role model, '... then he explicitly rejected this aspect of his style of management – the unwillingness to entrust responsibility to others.'[24] With his work completed on the GWR, Clark was to be employed on the TVR by Brunel in 1840.

Clark is credited with being the anonymous author of several books on the GWR during the time the main line was being built and after it was completed. This taste for writing appears to have manifested itself shortly after he qualified in medicine and had moved to Clifton. The most well known of these publications was The History and Description of the GWR, which included J.C. Bourne's famous lithographs and was published as a limited edition in 1843.[25] Clark's employment under Brunel appears to have ended, as with several other engineers, on the completion of the TVR, and he decided to change his career yet again and enter law, being admitted to the Middle Temple on 22 January 1842, although he was never called to the Bar. In 1843 Clark turned his back on opportunities in Britain to take up an engineering post in India. Henry Bartle Frere, the younger son of Edward Frere the ironmaster, (his second son, George Frere, has already been mentioned as one of Brunel's chief assistants on the GWR) was the Secretary to the Governor of Bombay. Bartle Frere and Clark became good friends and he suggested that there were opportunities for civil engineers in Bombay.[26] In India, Clark was to become involved with a survey for a railway out of Bombay over the Western Ghats, and on to the Deccan plateau. However, he turned down the post of engineer for the Great Indian Peninsula Railway Co. and returned home in 1847 to promote the proposed line and publish his report. The first part of the railway based on his plans, the first railway in India, was opened on 18 April 1853 as a 5ft 6in-gauge line from Bombay to Thana.[27]

What had brought him back to Britain, never to return to India, is unclear, but we do know that Guest may have had a hand in it as on 19 May 1845, Brunel wrote to Guest: 'In reply to your enquiries respecting our friend G T Clark I have great pleasure in bearing testimony to his high character as a man of honour and integrity I believe also that he has considerable talents…'[28]

Clark called to see the Guests on his return from India, visiting them in London: 'George Clark called and sat some time with us. He is not the least changed and seems as gay and buoyant as ever.'[29] This was two years on from that letter from Brunel, and any plans Guest may have had for Clark appear to have been put on hold temporarily, most likely due to the situation regarding the Dowlais lease. On the

East Moors steelworks, Cardiff, shortly after closure in 1978. (SKJ photograph)

same day that she recorded Clark's visit (13 May 1847), Lady Charlotte writes that hopes are quite gone of reaching a settlement for Dowlais. Consequently, Clark was to undertake work as a consulting engineer on his return home from Bombay, establishing himself in London and taking up employment as a superintending inspector under the Public Health Act from 1848.[30] Even though there was plenty of work to do for the Board of Health, visiting, inspecting and reporting on over forty towns and cities,[31] he retained his private practice. This work may have brought him back into contact with Brunel as there were plenty of opportunities for both to cross paths, for example Clark was engaged in surveys of towns through which Brunel was taking his railways. Clark's report map of Bridgend (drawn in 1848) shows the line and station of the South Wales Railway, then in the course of construction.[32] It is uncertain what contact he had with Brunel at this time, or after he submitted his final report to the Board of Health in April 1850, when he continued his private engineering practice.

Dowlais was now to figure largely in his future plans and it is more than likely that Clark met his future wife through the Guest circle. This was Ann Price Lewis, the sister of Wyndham William Lewis and the last of the Lewis family to be a partner in

John Calvert (1812-1890) and his Rhondda *steam engine in its original engine house. On completing his TVR contract, Calvert tried his hand at coal mining, sinking a pit at Gelliwion in the Rhondda in 1844, (the Newbridge colliery). Here he struck the celebrated No.3 Rhondda seam at 54 yards. The engine made by the Varteg Iron Co. (with a replacement cylinder by Brown Lenox) was to be used for pumping and winding in the Rhondda. It can be seen today in the grounds of the University of Glamorgan, Pontypridd (formerly a School of Mines). (SKJ collection)*

John Calvert's grave, Glyntaff churchyard, near Pontypridd. John Calvert was probably the most experienced railway contractor on the TVR, and had responsibility for a considerable section of the line, excluding major works. (SKJ photograph)

the Dowlais Iron Co.[33] Clark's engagement to Ann was announced at Canford Manor at Christmas in 1849, and they were married in Cardiff on 3 April 1850.[34]

By the time of Sir John Guest's death in 1852, Clark had become an established part of the Guest circle, and the Earl of Bessborough (grandson of Lady Charlotte), who was to edit Lady Charlotte Guest's journal in 1950, referred to him as '… Sir John's great friend and counsellor…'[35] Guest's will named Clark and Edward Divett as executors, along with Lady Charlotte, whose trusteeship of the Dowlais estate was to last for as long as she was a widow. Even before Guest's death she had thrown herself into managing the works, referring to the '… musical clank of bars of rail' and that the making of nail rods was, '… the prettiest process in the world.'[36] As well as seeking to understand the various ironworks processes, she was to become experienced in the wider business empire that Dowlais had become; after Guest's death she called in Nicholas Wood of Newcastle in an attempt to improve the level of coal extraction from Dowlais land. In June 1853, Anthony Hill had written to her to suggest a meeting of the four Merthyr ironworks to present a united front against impending wage claims and strike action by ironworkers.[37] By 1854, however, she had decided to remarry, an action that removed her from being directly involved in the management of Dowlais. Clark now became resident trustee and appointed professional managers such as William Menelaus, to take the Dowlais works in a new

Tylorstown collieries, Rhondda. Many generations living in South Wales today still have connections, if not direct experience, of working in the Rhondda. My wife's maternal grandfather made the change from agriculture to mining, being sent from Talybont-on-Usk to live with relatives at Tylorstown and to work in the pits. Here he was to follow his uncle by working at the Tylorstown collieries (Nos 8 and 9 pits), and his two sons would follow in his footsteps. The outbreak of the Second World War offered alternatives and, after active service, the two sons decided on a change of career, with one seeking opportunities in the USA. (Bill Daniels collection)

direction. Steel production was the answer to future prosperity, and Clark and his fellow trustee, Henry Austin Bruce (who was to replace Edward Divett as trustee), successfully negotiated with Henry Bessemer to adopt his process of steel production at Dowlais.[38] With the start of steel production in 1866, Dowlais began to import more and more of its iron ore as the process required a higher quality of hematite ore that could not be obtained locally. Ore was brought in from England and abroad, notably Spain, and had to be transported from the docks up to Merthyr, with the finished product sent back down the line for shipping. In the 1880s, Clark was to sanction one of the most far-reaching decisions to shift steel production from Dowlais to a new works at Cardiff beside the Roath and East Bute Docks. In some respects, this can be seen as the ultimate closure of the transport loop that had been initiated in shipping iron products to market at one end of the valley and supplying the process of ironmaking at the other. Dowlais was brought to the shipping place. Obviously there had been radical changes on the way; iron had given way to steel and the iron ores found at the heads of the valley were not suitable for the manufacture of steel. Coal for smelting could also be found in abundance at Cardiff as it was now the major dock export.

Iron had been overtaken by coal in terms of railway tonnages; the business opportunities of coal had been recognised at an early stage by the contractor John Calvert, who decided to abandon railway contracting and turn his hand to coal mining in 1844. In that year he entered into an agreement to work the coal under the property of Revd George Thomas and his brother, at Gelliwion. This undertaking became known as the Newbridge colliery where he struck the celebrated No.3 Rhondda seam at a depth of 54 yards. Here he installed a beam engine, one of the first applications of steam power for the purposes of winding in the Rhondda.[39] Four years later, in 1848, he sank a colliery at Gyfeillon but it was to take him three years to strike the No.3 Rhondda seam there, having to go down to 149 yards. Looking to supply coke for locomotive use he set up a battery of coke ovens at Gyfeillion, and sunk the Tymawr colliery. The result was a lucrative business with the GWR whereby Calvert sent large quantities of coke to Bristol.[40] In 1854 financial difficulties forced Calvert to sell the colliery to the GWR, following a three-month trial.[41] He later purchased the colliery, now known as the Great Western Colliery, back in 1864, and set up the Great Western Colliery Co. in 1866. By this time the use of coke for locomotive fireboxes had given way to burning bituminous coal.[42]

One of the most graphic indicators of the impact that the exploitation coal was to have on South Wales was the explosion in population of the Rhondda valley. By the end of the nineteenth century, the Rhondda had become the chief destination of the influx of population into Glamorgan, a movement which continued with increasing

Restored TVR-class 01 0-6-2T locomotive No.28. This was built by the TVR in May 1897 at the West Yard works at Cardiff Docks. It survived being sold out of service by the GWR in 1926 and eventually returned to South Wales in 1967. (Courtesy of the National Museum and Galleries of Wales)

The 1887 Queen Street station shortly before demolition in the early 1970s. (SKJ photograph)

vigour throughout the first decade of the new century.[43] Coal was not just a factor in the transformation of a railway's economic base but a transforming agent of society in Victorian South Wales. In 1841 there were 11,000 coal miners in South Wales; seventy years later, in 1911, the coal industry in South Wales was to employ 214,000. The industry in South Wales was to absorb more immigrants in the ten years up to 1911, faster than anywhere in the world excepting the USA.[44] Some 129,000 moved into the area, with the consequence that the majority of the population, at the start of the First World War, had been born outside the Rhondda.

From the outset only a short branch had extended into the Rhondda by the TVR. The largest customer in the Rhondda was Walter Coffin, who was happy to send his coal down his tramroad to the railhead at Dinas. The Taff Vale Railway Extension Act of 1846 sought to address this by extending the Rhondda branch, but there was a reluctance to proceed until the potential mineral wealth was confirmed. However, by March 1849 a new branch line between Eirw and Ynyshir was completed and, by May of the same year, it had reached Dinas. By this date the TVR had the necessary permission and land to construct a double line throughout the length of both valleys. With the progression of the TVR throughout the whole length of the Rhondda Fawr valleys, an expansive period in the history of the Rhondda began. Steam coal was proved in the Upper Rhondda Fawr in 1853, and the TVR was extended as far as Gelligaled by December 1855. It reached the Bute Merthyr colliery in Treherbert on 7 August 1856, although the first train of Rhondda steam coal had been sent down to Cardiff on 21 December 1855.[45] Within ten years, the Rhondda had become an attractive proposition as opinion regarded the existing coalfields, with shallower coal seams such as the Cynon valley, were being played out. The Rhondda's output in 1865 was only a quarter of that of the Cynon valley; in 1875 output had exceeded the Cynon valley's, with over 2 million tons.[46] Fifty years after that first trainload of Rhondda steam coal was sent down to Cardiff docks, Sir Charles M'Laren, chairman of the Tredegar Iron and Coal Co. summed up the position of Welsh coal in his article, 'The Welsh Coal Fields'. While admitting that the world leader in coal output was the United States with 320,000,000 tons in 1903 (the figure for Great Britain in 1904 was 232,306,147 tons), he affirmed:

> '... that British coal has never been surpassed in quality by the production of any foreign country, and it is beyond all dispute that the finest coal produced in these Islands, or in the world, is that which comes from South Wales.'[47]

This account was republished in 1906 by the *The Syren and Shipping*, the year that a Liberal Government came to power and passed the Trades Disputes Act in order to protect the interests of trade unions. This was an indirect consequence of action the TVR had taken in 1901, when it sued the Amalgamated Society of Railway Servants (ASRS) for losses it had suffered during a strike. The TVR won the case and the ASRS was fined £23,000. This became known as the Taff Vale Case and was a landmark ruling as up until then trade unions had assumed that they could not be sued for acts carried out by their members. Obviously it had significant

Taffs Well station footbridge with TVR device brackets in the 1970s. (SKJ photograph)

implications for trade unions as they could be sued every time they were involved in an industrial dispute. Despite political lobbying by the trades unions to redress this ruling, it was not until 1906, when the general election brought a Liberal administration into power, that legislation was taken to remove trade union liability for damage by strike action.

M'Laren was confident about the future of coal, and the high watermark as far as coal exports from South Wales were concerned was to come seven years later in 1913. The total output of coal from South Wales was some 57 million tons, Glamorgan accounting for 38 million tons, Cardiff docks exporting 13 million tons, and Barry docks, 11 million tons. The Barry Railway, led by the coal owner David Davies, was to tap into the markets that the TVR had considered to be in its franchise, particularly the Rhondda valleys. It was a railway that Henry Marc Brunel was involved with, but the incorporation of the Barry Railway in 1884 and the opening of the first Barry dock in 1889, caused the TVR to consider itself to be under threat. Despite the records being broken in 1913, there were greater threats to the South Wales railways in general and their trade in coal. There was a new contender for the fuel to drive industry, namely oil, the impact of which would be as disruptive as the global conflict that was to follow in 1914.

This view of the top of the main incline of the TVR shows a masonry bridge carrying a parish road across the cutting made when the tunnel was opened out. Compare this view with the photograph of the tunnel being opened out and the temporary timber bridge on p.146. The masonry bridge, with its deck railing and slender arch, has an ornamental quality and has long been replaced. From Ammon Beasley's interview in The Railway Magazine, *July 1898.*

Looking back; the wooden TVR platform shelter at Bute Road station, early 1970s. Like many Taff Vale Railway artefacts, it is now part of history. (SKJ collection)

The TVR was to become one of the most prosperous railways in Great Britain and is remembered today by the role it played in the formation of the South Wales valleys and by the prodigious amounts of coal it carried. Other achievements, namely the part it played as the successor to a railway locomotive system initiated by Richard Trevithick, the development of the Merthyr iron, and later steel, industry and its role as an urban commuter railway, which continues to this day, should not be overlooked. There is also the unique involvement of Brunel whose legacy in terms of engineering works is worthy of further examination.

'*When the Coal Comes From the Rhondda*

When the coal comes from the Rhondda
Down that Taff Vale Railway line
With my little pick and shovel
I'll be there!
I'll be there!'

Chapter 11 Notes

1 'When the Coal Comes From the Rhondda,' originated as a protest song in the folk tradition, around the Cambrian Combine lockout of 1910 which led to the Tonypandy Riots, and is now a popular rugby song.
2 Mountford, Eric R. and Kidner, R.W. (1995). *The AberdareRailway*, p.12, The Oakwood Press: Oxford.
3 Lloyd, John. (1906), *The Early History of the Old South Wales Ironworks 1760-1840*, pp.173-4, Bedford Press: London.
4 Ince, Laurence. (1993), *The South Wales Iron Industry, 1750-1885*, p.130, Ferric Publications: Solihull.
5 Mear, John F. (1999), *Aberdare, The Railways and Tramroads*, pp. 75-6, published by the author: Aberdare.
6 Lloyd, John. (1906), pp.111-12.
7 Jeffreys Jones, T I. ed. (1966), p.87, *Acts of Parliament Concerning Wales 1714-1901*. Cardiff: University of Wales Press.
8 Mountford, Eric R. and Kidner, R.W. (1995), p.12.
9 Mountford, Eric R. and Kidner, R.W. (1995), p.13.
10 Mountford, Eric R. and Kidner, R.W. (1995), p.13.
11 *Glamorgan, Monmouth and Brecon Gazette, Cardiff Advertiser and Merthyr Guardian*, 22 December 1846.
12 Mountford, Eric R. and Kidner, R.W. (1995), pp.28-9.
13 Williams, John, (1980). *The Coal Industry, 1750-1914*, pp.189-90. Chapter IV in *Glamorgan County History, Volume V, Industrial Glamorgan from 1700 to 1970*. Edited by John, Arthur H., and Williams, Glanmor, Glamorgan County History Trust Limited/University of Wales Press: Cardiff.
14 Wilkins, Charles, (1888), *The South Wales Coal Trade and its Allied Industries from the earliest days to the present time*, p.191, Daniel Owen & Co. Ltd: Cardiff.
15 Wilkins, Charles, (1888), p.190.
16 Mountford, Eric R. and Sprinks, Neil. (1993). *The Taff Vale Lines to Penarth*, p.7, The Oakwood Press: Oxford.
17 Mountford, Eric R. and Sprinks, Neil. (1993), p.9. The PHDR's third Act was obtained on 22nd June 1863 and authorised the lease period of 999 years. The lease for the tidal harbour and railway from Penarth junction to the harbour was to take effect from 1st January 1864.
18 Brunel Archives, Brunel University. Letter by H.M. Brunel outlining professional career to date; February 1st 1877. H M Brunel Collection.
19 Brunel Archives, Brunel University. Letter by H.M. Brunel outlining professional career to date; February 1st 1877. H M Brunel Collection.
20 Network Rail, Bridge over Newport T.P. Road (Adam Street, Cardiff), signed and dated John Hawkshaw, 6 October 1858.
21 Williams, Caroline E. (1893), *A Welsh Family*, Women's Printing Society Ltd: London.
22 James, Brian Ll, Ed, (1998), *G.T. Clark, Scholar Ironmaster in the Victorian Age*, The Making of a Scholar Ironmaster, p.8, University of Wales Press, Cardiff.
23 Brunel, Isambard, (1870, reprinted 1971), *The Life of Isambard Kingdom Brunel*, p.94-5, Longmans, Green & Co., London (reprinted David & Charles, Newton Abbot).
24 James, Brian Ll, Ed, (1998), p.9.
25 James, Brian Ll, Ed, (1998), p.11.

26 James, Brian Ll, Ed, (1998), p.12.
27 Berridge, P. S. A. (1969), *Couplings to the Khyber*, p.17, David & Charles: Newton Abbot.
28 Brunel Archives, Brunel University, Letter Book 4, p.8, Brunel to J J Guest, 19 May 1845.
29 Bessborough, The Earl of,ed, (1950), p. 186, *Lady Charlotte Guest Extracts from her Journal 1833-1852*, John Murray: London.
30 'G.T. Clark, Slums and Sanitary Reform' by Andy Croll in James, Brian Ll, Ed, (1998), *G.T. Clark, Scholar Ironmaster in the Victorian Age*, University of Wales Press: Cardiff.
31 James, Brian Ll, Ed, (1998), p.26.
32 Randall, Henry John. (new edition 1994), *Bridgend: the Story of a Market Town*, p.128, Cedric Chivers: Bristol for Mid Glamorgan County Libraries.
33 James, Brian Ll, Ed, (1998), p. 4.
34 Bessborough, The Earl of,ed, (1950), p.233.
35 Bessborough, The Earl of,ed, (1950), p.233.
36 Guest, Revel and John, Angela V. (1989), *Lady Charlotte ; A Biography of the Nineteenth Century*, pp.120-21, Weidenfeld and Nicolson: London. See also the chapter on; 'The Head of the Works'.
37 Guest, Revel and John, Angela V. (1989), p. 172-74.
38 James, Brian Ll, Ed, (1998), p.60.
39 Cornwell, John. (1983), *The Great Western and Lewis Merthyr Collieries*, p.4, D. Brown & Sons Ltd: Cowbridge. Calvert's beam engine was made by the Varteg Ironworks about 1845 (with a replacement cylinder made in 1861 by Brown Lenox). In 1918 it was presented to the School of Mines at Treforest, now the University of Glamorgan, where it can be seen today.
40 Powell, Don. (1996), p.122, *Victorian Pontypridd and its villages*, Merton Priory Press: Cardiff.
41 MacDermot, E T, revised by Clinker, C R. (1964), p.402. *History of the Great Western Railway*, Vol.1. London: Ian Allen. MacDermot puts the acquisition down to, '...the failure of a contractor for the supply of the Bristol coke ovens; it was worked by the Company till 1865 and then sold.'
42 MacDermot, E.T. (1964), p. 402. '...it was not until 1857 that coal finally replaced coke in locomotives with ordinary fireboxes.'
43 Lewis, E.D. (Second edition 1963), *The Rhondda Valleys*, p.231, Pheonix House: London.
44 Royal Commission on the Ancient and Historic Monuments of Wales (RCAHM). ( ), *Collieries of Wales Engineering & Architecture*, pp.148-49, RCAHM: Aberystwyth.
45 Lewis, E.D. (1963), p.69.
46 Williams, John, (1980). p.182. Chapter IV in *Glamorgan County History, Volume V.*
47 *The Welsh Coal Fields*, p.7, the opening chapter by Sir Charles M'Laren appears as a series of articles originally published in *The Syren and Shipping* and repinted in 1906, The Syren and Shipping Ltd: London.

INDEX

Bold italics denotes illustrations

GENERAL INDEX

INDEX OF PEOPLE

If you are interested in purchasing
other books published by Tempus, or in case you have
difficulty finding any Tempus books in your local bookshop,
you can also place orders directly through our website

www.tempus-publishing.com